Helion & Company Limited
Unit 8 Amherst Business Centre
Budbrooke Road
Warwick
CV34 5WE
England
Tel. 01926 499 619
Email: info@helion.co.uk
Website: www.helion.co.uk
Twitter: @helionbooks
Visit our blog http://blog.helion.co.uk/

Designed and typeset by Farr out Publications, Wokingham, Berkshire
Cover design Paul Hewitt, Battlefield Design (www.battlefield-design.co.uk)

ISBN 978-1-915070-67-8

British Library Cataloguing-in-Publication Data
A catalogue record for this book is available from the British Library

We always welcome receiving book proposals from prospective authors.

CONTENTS

MAP OF EUROPE 1945–1992

Note: In order to simplify the use of this book, all names, locations and geographic designations are as provided in *The Times World Atlas*, or other traditionally accepted major sources of reference, as at the time of described events.

SPANISH AIR FORCE ACRONYMS		
EJÉRCITO DEL AIRE Y DEL ESPACIO		
ACGEA	Agrupación del Cuartel General del Ejército del Aire	EdA's Headquarters Group
ABA	Academia Básica del Aire	Basic Air Academy
AB		Air Base
ACAR	Acuartelamiento Aéreo	Air Quartering
AGA	Academia General del Aire	General Air Academy
AIS		Automatic Identification System
AJA	Asesoría Jurídica del Aire	Air Legal Advisory
CASYD	Centro de Adiestramiento de Seguridad y Defensa	Defence and Security Training Centre
CAOC-TJ	Centro de Operaciones Aéreas Combinadas	Combined Air Operations Centre – Torrejón
CECAF	Centro Cartográfico y Fotográfico	Cartographic and Photographic Centre
CESAEROB	Centro de Sistemas Aeroespaciales de Observación	Centre for Aerospace Observation Systems
CIGES	Centro de Informática de Gestión	Centre for Management Informatics
CIMA	Centro de Instrucción de Medicina Aeroespacial	Centre for Aerospace Medicine Education
CLAEX	Centro Logístico de Armamento y Experimentación	Weapons and Experimentation Logistics Centre
CLOIN	Centro Logístico de Intendencia	Logistical Supply Centre
CLOMA	Centro Logístico de Material de Apoyo	Logistics Centre for Support Material
CLOTRA	Centro Logístico de Transmisiones	Transmission Logistics Centre
CMOPS	Comandante del Mando de Operaciones	Operations Command Commander
CN-ETAC	Centro Nacional-European Tactical Airlift Centre	National Centre – European Tactical Airlift Centre
COVE	Centro de Operaciones de Vigilancia Espacial	Space Surveillance Operations Centre
DAE	Dirección de Asuntos Económicos	Economic Affairs Directorate
DLO	División de Logística	Logistics Division
DOP	División de Operaciones	Operations Division
DPL	División de Planes	Plans Division
DISAN	Dirección de Sanidad	Directorate of Health
EdA	Ejército del Aire	Spanish Air Force
EADA	Escuadrón de Apoyo al Despliegue Aéreo	Air Deployment Support Squadron
EADS		European Aeronautic Defence and Space
ECAO	Escuadrilla de Circulación Aérea Operativa	Operational Air Circulation Squadron
EMA	Estado Mayor del Aire	Air Force Staff
EMACON	Estado Mayor Conjunto de la Defensa	Joint Defence Staff
EMACOT	Escuela de Técnicas de Mando, Control y Telecomunicaciones	School of Command, Control and Telecommunications Techniques
ESAOGEL	Escuadrón de Apoyo Operativo a la Guerra Electrónica	Electronic Warfare Operational Support Squadron
ESAPO	Escuela de Apoyo	Support School
ESRO		European Space Research Organisation
ESTAER	Escuela de Técnicas Aeronáuticas	School of Aeronautical Techniques
ESTRAM	Escuadrilla de Transmisiones	Transmission Squadron
ETESDA	Escuela de Técnicas de Seguridad, Defensa y Apoyo	School of Security, Defence and Support Techniques
EVA	Escuadrón de Vigilancia Aérea	Air Surveillance Squadron
EZAPAC	Escuadrón de Zapadores Paracaidistas	Paratrooper Sappers Squadron
FLIR		Forward Looking Infra Red

GABJEMA	Gabinete del Jefe del Estado Mayor del Ejército del Aire	Cabinet of the EdA's Chief of Staff
GRUALERCON	Grupo de Alerta y Control	Alert and Control Group
GRUCAO	Grupo de Circulación Aérea Operativa	Operational Air Circulation Group
GRUCEMAC	Grupo Central de Mando y Control	Central Command and Control Group
GRUEMA	Grupo Escuelas de Matacán	Matacán Schools Group
GRUMOCA	Grupo Móvil de Control Aéreo	Mobile Air Control Group
GRUNOMAC	Grupo Norte de Mando y Control	Northern Command and Control Group
GRUTRA	Grupo de Transmisiones	Transmission Group
IDC	Intervención Delegada Central	Audit Office
JEMA	Jefe del Estado Mayor del Aire	Chief of the Air Staff
JEMAD	Jefe del Estado Mayor de la Defensa	Chief of Defence Staff
JMOVA	Jefatura de Movilidad Aérea	Air Mobility Headquarters
JSMC	Jefatura del Sistema de Mando y Control	Command and Control System Headquarters
JSTCIS	Jefatura de Servicios Técnicos y de Sistemas de Información y Telecomunicaciones	Head of Technical Services and Information and Telecommunications Systems
LRR		Long Range Recon
MACAN	Mando Aéreo de Canarias	Canary Islands Air Command
MACOM	Mando Aéreo de Combate	Combat Air Command
MAESAL	Maestranza Aérea de Albacete	Albacete Air Workshops
MAESE	Maestranza Aérea de Sevilla	Seville Air Workshops
MAESMA	Maestranza Aérea de Madrid	Madrid Air Workshops
MAGEN	Mando Aéreo General	General Air Command
MALOG	Mando de Apoyo Logístico	Logistic Support Command
MAPER	Mando de Personal	Personnel Command
MDOA	Mando de Defensa y Operaciones Aéreas	Defence and Air Operations Command
MOE	Mando de Operaciones Especiales	Special Operations Command
MOPS	Mando de Operaciones	Operations Command
NBC		Nuclear Biological Chemical
PMJEMA	Puesto de Mando del Jefe del Estado Mayor del Ejército del Aire	Command Post of the EdA's Chief of Staff
SAM		Surface to Air Missile
SAR	Servicio Aéreo de Rescate	Rescue Air Service
SEADA	Segundo Escuadrón de Apoyo al Despliegue Aéreo	Second Air Deployment Support Squadron
SEJEMA	Segundo Jefe del Estado Mayor del Ejército del Aire	Second Chief of the EdA's Staff
SHYCEA	Servicio Histórico y Cultural del Ejército del Aire	EdA's Historical and Cultural Service
STOL		Short Take-Off and Landing
USAF		United States Air Force
UAS	Escuela de Sistemas de Vuelo no Tripulado	Unmanned Flight Systems School
UAV		Unmanned Aerial Vehicle
UMAAD	Unidad Médica Aérea de Apoyo al Despliegue	Aerial Medical Unit for Deployment Support
UMAER	Unidad Médica de Aeroevacuación	Aero Evacuation Medical Unit
UME	Unidad Militar de Emergencias	Military Emergency Unit
COMPANIES		
AISA	Aeronáutica Industrial, S.A.	Spanish aircraft manufacturer
CASA	Construcciones Aeronáuticas, S.A.	Spanish aircraft manufacturer

CETA	Compañía Española de Técnica Aeronáutica, S.A.	Engineering company with experience in aircraft modernisation
ENHASA	Empresa Nacional de Hélices de Aviación, S.A.	Spanish propellers company
ENMASA	Empresa Nacional de Motores de Aviacion, S.A.	Spanish aircraft engine company
EXPAL	Explosivos Alaveses, S.A.	Munitions and explosives factory.
HASA	La Hispano Aviación, S.A.	Former Spanish fighter aircraft factory
INTA	Instituto Nacional de Técnica Aeroespacial	National Institute for Aerospace Technology
PROGRAMMES		
ALAD	Arma Lanzada a Distancia	Remote Launched Weapon
FACA	Futuro Avión de Caza y Ataque	Future Fighter and Attack Aircraft
FALES	Futuro Avión Ligero de Enseñanza Selectiva	Future Selective Training Light Aircraft
FATAM	Futuro Avión de Transporte Aéreo Medio	Future Medium Air Transport Aircraft
MODAR	Modernización del Armamento	Armament Modernization
POBAS	Potenciación de las Bases Aéreas	Enhancement of Air Bases

INTRODUCTION

What the reader has in his hands is intended to be a summary of what makes up the Spanish Air Force (Ejército del Aire, EdA) at the present time.

The author has been following as closely as possible, the development of the Spanish Air Force from 1975 to the present day. I have witnessed, from a journalistic and historical point of view, the evolution that this force has undergone over the course of these more than 45 years. I must confess that it has not been an easy task since, under the cover of discretion, the main source, which is the Force itself, has almost never facilitated the collection of data. The only information that this official authority has provided has always been very general, avoiding providing details. The sources that I have used during these decades have almost always been indirect, through third parties or simply analysed in depth what was coming to me.

In recent years and with the generational change, this situation has improved a little and, fortunately, at the present time there has already been a limited opening, which has facilitated this type of work.

The content of this book is a summary of what the author has collected over several decades. The book is intended to be a reference manual for those interested in this Air Force. I have left for another occasion to describe in much more detail and depth the history of this Force from its creation to the present day. During these long years, air bases and military aerodromes have disappeared, both for economic reasons and for logistical and organisational reasons. The same can be said of aircraft and helicopters, although in this case it has been to modernise air resources as result of technical advances. The personnel, both pilots and ground crew, have also evolved and with it, the way of seeing things and acquiring new knowledge and new ways of using the technological means at their disposal. It should be mentioned that many of the aircraft at the disposal of the EdA, particularly transport aircraft and helicopters, have changed their function and changed units, and on many occasions, for reasons of discretion, it has not been easy to monitor them. What the reader will find in the lists of aircraft in service represents the most current data.

In addition to combat units, the EdA has several facilities that serve and support flight units and personnel. All this is reflected in this work.

The EdA is fully integrated into NATO and is responsible not only for defending Spanish territory, including the two archipelagos of the Balearic and Canary Islands, but also southern Europe, together with the Air Forces of the countries in the area.

This has not been, nor is it, an easy task for the Spanish State, both from an economic and a physical and human point of view. For years, the Spanish Armed Forces have been misunderstood and the investments the State makes in them, are not always appropriate. They have one of the lowest budgets in Europe, which means that flight equipment ages and is difficult to replace so this means that ground staff must make great efforts to keep it operational.

Year after year, the Spanish Armed Forces (EdA, Army, and Navy) have seen their investments reduced, resulting in a loss of capabilities which, despite the commitments made by politicians to improve them, have never been resolved and the deterrent power exercised has therefore, decreased alarmingly. In recent decades, the EdA has suffered a drastic reduction in the level of investment. We can say that in the penultimate economic crisis that Spain has suffered (between 2008 and 2014), budget cuts amounted to a third, and this occurred in the context of significant payments (Eurofighter, A 400, and others). The loss of a third of the investment is always a hard knock, given the insensitivity shown by the leaders of the Ministry of Defence, of all political trends, who give the impression that security and defence issues are not especially important to them. This is in despite of the increase in tensions and global geostrategic approaches, which are particularly relevant to our immediate environment (North Africa) and nearby areas of influence (western Mediterranean, Strait of Gibraltar, and others).

Despite the appreciation that international organisations have for the Spanish Armed Forces (as evidenced by the constant invitations to take part in international missions), the critical statements relating to the lack of a unified and stable foreign policy are well known and should make it possible to draw up a clear strategy. This circumstance results in Spain's scant political and strategic weight in the world. As a result, our country is not an important player in the international environment, depending almost always on third parties to achieve certain objectives. This was demonstrated by the Covid-19 health crisis (Spring 2020), with the problem initially presented by attempts to procure vital health materials to deal with the pandemic.

The lack of definition of Spain's role in the world prevents the selection and achievement of objectives, which in the case of the EdA, means the modernisation of facilities and flight equipment.

The lack of serious strategic planning leads to a degradation of medium and long-term objectives. Aeronautical programmes are becoming increasingly complex and expensive and therefore, decisions must be taken well in advance. A clear example is that at the time of writing, the EdA (and also the Navy's Air Force, Arma Aérea de la Armada, AAA) should be making serious plans to replace the now obsolete McDonnell-Douglas/Boeing F-18 (and AAA's AV-8B Harrier IIs) with the only thing on the market, the Lockheed-Martin F-35 Lightning II. It is feared that the decision will be taken at the last moment and in an ill-considered manner, which will mean that the number of units contracted will not be maintained. Also, the reduction in the number of units purchased will be drastically reduced at the time of production, a similar issue happened with the F-18, Eurofighter and A 400, where the original orders were reduced by almost a third.

We cannot end this Introduction without mentioning that during 2019, this Air Force performed a total of 65,000 flight hours with an average availability of 54 percent of the aircraft crews, data that highlights the shortcomings mentioned in the previous paragraphs.

1
BRIEF HISTORY OF THE SPANISH AIR FORCE (EJÉRCITO DEL AIRE, EdA), 1939-2022

THE EARLY TIMES

The first military aeronautical activity is dated in 1911 and located at the Cuatro Vientos Aerodrome (Madrid), in which an aviation school and an experimentation centre were established for the purpose of its military application. However, the reader must go back to the end of the previous century and look for the roots of flight in the aerostatic activities that were developed for observation purposes to support the men of the Army.

The Military Aeronautics Service (Servicio de Aeronáutica Militar) was created on the eve of the Great European War, on 28 February 1913. It brought together, under a single command, the aerostation and the incipient aviation. The latter branch was equipped with five Nieuports suitable for war services, aircraft which would later be complemented by the Farman MF.7 and the Lohner Pfeilflieger, all of them front-line aircraft, similar to those fighting in northern and central Europe.

During that same year, some of these aircraft were moved to northern Morocco, a territory where Spanish troops disputed their stay there. Equipped with very primitive bombs and scopes for their launch they were successful in supporting the troops.

In 1918, the Aeronautic of War Section (Sección de Aeronáutica de Guerra) was set up and a good number of aircraft were purchased from the war surplus of the victorious Allies.

Air bases were set up to house and organise new units, and calls were issued to cover pilot positions. On an industrial level, competitions were held for the development of Spanish aircraft.

Between 1920 and 1921, courses were established in five aviation schools. And a new training course was created called 'Aviation Complement'.

In 1922, the foundations were laid for the reorganisation of the Military Aeronautics (Aeronáutica Militar), whilst in 1926, the Superior Headquarters of Aeronautics (Jefatura Superior de Aeronáutica) was created.

All this enthusiasm made the whole country participate, in one way or another, in the development of this new way of flying and fighting. It was then that the Great Flights emerged: Madrid-Tetouan (1919), Iberian Peninsula-Canary Islands (1924), Huelva-

Hispano Aviación, HA.1112K Buchón. (Author's Photo)

Buenos Aires (1926), Raid to the Philippines (1926), Atlantic Patrol to Spanish Guinea (1926), Seville-Mesopotamia ('Jesús del Gran Poder', 1928) and Seville-Brazil ('Jesús del Gran Poder', 1929). The proclamation of the Republic did not cut short these initiatives so we see the Seville-Cuba Raid (1933) and the Mediterranean circuit of Los Alcázares and Mar Chica seaplanes (1933 and 1934).

The outbreak of the Civil War cut off these initiatives, the pilots now had other, less spectacular, but more risky missions.

1939–1952

The Spanish Air Force (Ejército del Aire, EdA) was born between 1936 and 1939 without any official provision to support its creation. For this reason, the Law of 7 October 1938 referred in its preamble, of the need to establish a legal text with which to consolidate its existence.

Once the Civil War was over, Spanish aviation focused on keeping the aircraft it had at the end of the war in the air and becoming familiar with some 300 combat and transport aircraft. It was then that an aerial demonstration was held in Barajas in which 451 aircraft took part.

Professional aviators continued to rise through their original Corps and Army, but the Law of 9 November 1939, created the Aviation Authority (Arma de Aviación), which they all joined. Two days later and by means of a new order, the Air and Ground Scales (Escalas de Aire y Tierra) were created, through which the specialities were defined.

On the 22nd of the same month, the Aviation Academy (Academia del Arma de Aviación) was founded, from which the first non-commissioned officers of the new army would emerge.

The period 1939–1942 saw the emergence of a whole infrastructure that replaced that improvised one during the war years. It was then that land was acquired on which to build the main air bases and barracks.

As far as aircraft are concerned, a selection of different types of aircraft were made during these years. At the same time, preparations were made for the production, under licence, of the fighter Me.109, the bomber He.111, the transport Ju.52 and the trainer Bü.131.

Four hundred Me.109s and He.111s were produced at a good pace but could not be finished on time due to a lack of engines from abroad. The outcome of the Second World War prevented this. In the meantime, the aircraft received by both sides during the last months of the Civil War was completed.

During the Russian campaign, an EdA's Squadron participated (1941–1944) in the air combat on the Eastern Front. This intervention aggravated the isolation to which Spain was subjected, which from the aeronautical point of view made the acquisition of aircraft and spare parts even more difficult.

In 1941, General Instruction No. 1 changed the numerical system of aircraft codes, which helped to clarify aircraft availability.

The Spanish aeronautical industry began developing engines from those available at that time. This is how the well known 'Beta', 'Tigre' and others were born. This activity also benefited education, as the Elizalde firm created its own school.

To be able to fly the Me.109s and the He.111s, manufactured by La Hispano Aviación, S.A. (HASA) and by Construcciones Aeronáuticas, S.A. (CASA), several batches of surplus engines from the Second World War were bought from France. The result was not particularly good, but it allowed production to be completed. These engines were later replaced by the Rolls Royce 'Merlin' of British origin.

In 1953 the form of coding was changed for some aircraft, mainly those related to school tasks.

In 1945, the General Air Academy (Academia General del Aire, AGA) began to operate. The first promotion came out in 1949.

From 1946 onwards, a good number of flight units and ground personnel appeared. Examples include the School of Motoring and Transmissions (Escuela de Automovilismo y Transmisiones), the Photographic and Cartographic Centre (Centro Fotográfico y Cartográfico), and several others which complemented those already established.

1953–1969

The first and timid contacts with the United States took place in April 1951, when several USAF pilots visited Spain and held exhibitions in Madrid and Seville. Shortly afterwards, the Spanish air minister visited the Philippines and several US air bases there. These contacts culminated in the signing in September 1953, of the Friendship and Cooperation Agreements between Spain and the United States of America, through which the Spanish Armed Forces (and in this case, the EdA) received a good number of aircraft and other material with which to update the obsolete military aviation.

Following the agreements, the three air bases for joint use were defined: Torrejón, Zaragoza and Morón (and Rota for the US Navy). Construction began immediately and after several stages, modern facilities with all kinds of equipment were available.

Lockheed T-33A, E.15-36, 41-24. (Author's Photo)

North American T-6D Texan, C.5-135, 514-5. (Author's Photo)

North American F-86F Sabre, C.5-98. (Author's Photo)

AISA I-115 Garrapata, E.9-46, 371-3. (Author's Photo)

Boeing KC-97 Stratotanker, TK.1-2, 123-02. (Author's Photo)

On 24 March 1954, the first six Lockheed T-33A training jets and a group of instructors arrived in Talavera La Real.

In July, the first North American T-6 Texans arrived for Matacán.

The first fighter jet pilots were trained in the United States and the Federal Republic of Germany, from which the first North American F 86 Sabres arrived.

In April 1956, 1st Wing (Ala Núm. 1) was formed in Manises, the first with jets. Later the 41st Squadron (41 Escuadrón) – later named 4th Wing – was created.

In May 1954, the first two Grumman HU-16 Albatross were received and shortly afterwards, the 50th (Search and Rescue) SAR Squadron (50 Escuadrilla del SAR) was organised.

These years were a period of profound reorganisation in which various units disappeared and many others were created equipped with modern flight equipment. The Me.109s and He.111s, given their recent incorporation, shared the skies of Spain with the jets coming from beyond the Atlantic.

The industry, for its part, also obtained important results.

In 1955 the first Saetas, Azores and Dorniers Do.27 flew, as well as the Aerotécnica AC.13 helicopters. However, the massive influx of foreign material ended these initiatives and forced the industry to concentrate on reviewing foreign material.

In April 1956, the Air Defence Command (Mando de la Defensa Aérea) was set up, incorporating all the appropriate units and with the means to deal with any external threat.

The events in Ifni and the Sahara did not take the EdA by surprise and in anticipation of what might happen, it sent a good number of CASA 2.111s and HA 1.109s to the Canary

CASA C.207 Azor, T.7-20, 405-20. (Author's Photo)

HA.220D Saeta, C.10B-73, 406-73. (Author's Photo)

CASA C.127, L.9-4, 903-04. (Author's Photo)

Islands, which intervened in repelling the aggression when the time came. Although the campaign was limited to the winter of 1957–58, the presence of important air units in the area was extended until they abandoned the territory several decades later.

The sixties saw the presence of the legendary F-104 Starfighters, which, despite their small number, had a great influence on the way air combat was conceived. It was also during this period, that the decision was taken to incorporate a good batch of Northrop F-5s, which, assembled by CASA, pushed the national industry to achieve greater goals.

If combat aviation achieved a very acceptable trajectory at that stage, the same was not true of transport aviation, which continued to be anchored in the past with its CASA 352s and the Douglas DC-3s. This was not always going to be the case, as in 1968, the EdA and CASA joined forces to develop a small transport that would become a great sales success over the years.

1970–1999

From an organisational point of view, this period began with the modernisation of the military structure of the Army. In 1971 the Directorate of Plans and Programmes (Dirección de Planes y Programas) was created, which was the institution designed to carry out the task.

Along with this update, came the renewal of the agreements with the USA and the implementation of the programme to incorporate 30 Dassault Mirage IIIs (bought in 1968) of French origin. The latter was due to the need not to depend on a single supplier, as the agreements with the United States limited the use of the material from their origin in the event of a conflict.

At the beginning of this period, the F-86s that for so many years were the basis of combat aviation, were already in a phase of withdrawal. With them disappeared the 101st and 102nd Squadrons, which were replaced by the 11th and 12th Wings equipped with the new Mirage III and the powerful McDonnell-Douglas F-4C Phantoms II. On the other hand, F-104 Starfighters were returned to the USA in 1972. At the same time, the 202nd Squadron began to receive the Northrop F-5s, a unit which, when upgraded, became 21st Wing.

The Villanubla AB witnessed the empowerment of the Saeta, who, suitably armed, switched to tactical aviation.

The year 1972 saw the end of the CASA 2.111s (the He.111s manufactured in Spain), whilst the newly formed 221st Squadron of the 22nd Wing received the first Lockheed P-3B Orions.

Finally, transport aviation upgraded its units to Wings. The former 35th and 37th Groups were strengthened with the incorporation of the CASA Azores and the DeHavilland Caribous. In 1973, the 301st Squadron was formed, which would soon receive the first C-130 Hercules. At that time, the first C.212 Aviocar units also entered service, which allowed the old CASA 352s (Ju.52) to be written off.

Avions Marcel Dassault Mirage IIIDE, C.11-2, 112-1. (Author's Photo)

Lockheed KC-130H Hercules, TK.10-06, 301-06 (31-51). (Author's Photo)

Avions Marcel Dassault Mystère 20D, TM.11-03, 401-03 (47-23). (Author's Photo)

Squadrons in El Aaiun. The 301st Squadron also specialised in transport flights between Gando and El Aaiun.

On an industrial level, in 1972 CASA bought Hispano Aviación.

In 1978, Directive 1/78-DOR (ORGEA Plan) was made public, aimed at thoroughly remodelling the EdA, always in search of greater operability. At that time, there were 617 aircraft and 77 helicopters.

Between 1979 and 1982, three Fokker F 27 were added for the SAR and 70 CASA C.101 Aviojet for school and training. Constant evolution meant that aircraft loaded with flying hours disappeared, giving way to modern aircraft designed to reinforce the units.

At the end of the 1970s, the idea was to replace the Phantom II with a new aircraft that would meet the needs of combat aviation well into the twenty-first century. The FACA Programme (Futuro Avión de Caza y Ataque, Future Fighter and Attack Aircraft) covered the range of fighter and attack aircraft that were beginning to be used at that time, as well as those that were under development.

In 1983, the decision was taken to purchase 144 F-18A Hornet twin-engined fighter-bombers, a number that was eventually reduced to 72.

The deliveries began in 1986 and the 12th and 15th Wings were equipped with them.

At the Talavera Fighter School, the T-33s were replaced by the modern F-5s, although the former would continue to operate in the retraining 41st Group.

In the long chapter on material losses, we will highlight the withdrawal of the last seaplanes from the EdA. These were the legendary Dornier Do 24s that joined in the 1940s and carried out hundreds of SAR missions from Pollensa AB.

In 1975, the first transport helicopters were purchased, the Aerospatiale SA.330H Puma with which, a special unit based at Cuatro Vientos was created.

Once again, the Sahara was an integral part of the history of Spanish aviation. In 1974 and with the action of the POLISARIO Front, instability returned to the territory. On this occasion, the Gando's 46th Mixed Wing deployed part of its 462nd and 463rd

The FATAM Programme (Futuro Avión de Transporte Aéreo Medio, Future Medium Air Transport Aircraft), which was intended to promote transport aviation, took years to implement but was finally taken forward with the incorporation of twenty CASA CN.235.

At the same time, other programmes were also started which, although less well known, were also of great importance. The main ones were aimed at modernising both armaments and air bases (MODAR, Modernización del Armamento, and POBAS, Potenciación de las Bases Aéreas).

Through these programmes, highly specialised missiles were acquired (Harpoon, Harm, et cetera) and armoured shelters were built in Gando and Albacete bases.

It has been left it to the end to mention the incorporation of the Dassault Mirage F.1s, since, apart from the aircraft originally purchased from the manufacturer, during the 1970s more aircraft were acquired from L'Armée de l'Air and the Qatar Air Force.

At the end of the 1990s, the star programme was the EFA (European Fighter Aircraft), whose aircraft developed in Europe were due to enter service within a few years.

2000–2020

At the beginning of the 2000s, the EdA was developing different aeronautical programmes to maintain its potential. The major European projects were related to the EFA Programme and the Airbus A.400 Atlas. At the national level, the FATAM Programme (CASA C.295) should be mentioned, which served to provide the EdA with a transport aircraft to cover multiple missions.

Within the framework of helicopters for SAR missions, transport, and school and training, the aim was to modernise the flight equipment, thus incorporating the Airbus H-215 (AS.332 C1) Super Puma and Eurocopter AS.532 UL Cougar. The Eurocopter EC.120 Colibri was purchased for the teaching environment. Recently, the first Airbus NH.90 Lobo has started to be supplied.

In 2003, deliveries of the first – out of a total of 78 – Eurofighter EF.2000 Typhoons began, intended to replace a good number of F-18 Hornets of the 12th, 15th and 46th Wings, whilst in 2016 the first of fourteen Airbus A.400 Atlas arrived to replace the veteran C-130 Hercules of the 31st Wing.

The objective of renewing the training aircraft fleet was achieved at the beginning of 2020 with the purchase contract for twenty-four Pilatus PC-21, destined for the AGA.

In relation to UAVs, four General Atomics Predator B/MQ-9 Reapers (with all ground elements) have been incorporated in 2020. It also has a certain number of

HA-200-R-1. This is the second of the two prototypes produced in 1956 and first flew on 11 January 1957. With the civil registration EC-AMM it was shown at the Le Bourget Salon in 1957. In the foreground is one of the Marboré engines. It is in the Air Museum of Cuatro Vientos (Madrid). (Author's Photo)

In a workshop is seen the original C-10C-92, 214-92, which later became A-10C. This Super Saeta carries on its tail the 214th Squadron's shield. (Author's Photo)

This is the A.10C-104, 214-104. It is armed with two containers carrying two Browning M-3 12.7mm machine guns and an ammunition box capable of 250 rounds and four 70mm rocket launchers. Note that there is no unit code on the fuselage. (Author's Photo)

This photo is from the 1973 Le Bouget Salon. Seen is the author of this book next to the HA.220 Super Saeta. (Author's Photo)

AeroVironment Wasp RQ-12A and RQ-11 Raven with which to cover current needs, pending several developments of completely Spanish aircraft being tested.

2021 ONWARDS

At the beginning of the current decade, EdA is involved in a series of programmes aimed at modernising its air power. In summary, it has proposed to acquire four Airbus/CASA C.295 ASW for maritime patrol (SAR) and naval warfare (ASW) missions.

Together with the Luftwaffe, it is proposed to equip the future Eurofighter Typhoon of the Halcón Programme with the new AESA (Active Electronically Scanned Array) CAPTOR-E radars. The EdA will equip its 20 new Typhoons of the 46th Mixed Wing in Gando (Canary Islands) with 20 of them.

Three Airbus A.330s, which will eventually be converted into A.330MRTTs (Multi Role Tanker Transport), have been purchased from the Spanish civil airline Iberia.

In the coming years, it will continue to be equipped with Airbus Helicopters H.135. The programme is expected to end in 2026.

It will initiate a programme to acquire NGWS/FCAS (New Generation Weapons System/Future Combat Aircraft System) aircraft. This is a tri-national programme also involving France and Germany.

Development of a joint study programme with France and Germany to produce an advanced training aircraft for the future NGWS.

Study related to the possible participation in the MALE Eurodrone Programme (Medium Altitude Long Endurance).

Development of the Halcón Programme to replace the F-18 Hornets from the US Navy, currently in service in the 46th Mixed Wing, which are at the end of their operational life. This Programme provides for the purchase of twenty Eurofighter Typhon from Tranche 3+/4.

Airbus, at the request of the Directorate General for Armaments and Material (DGAM), is proposing an advanced training aircraft to replace in the short term, the F-5Ms of the 23rd Wing, and the C.101 Aviojets of the AGA. The proposed model is called the Advanced Fighter Jet Trainer (AFJT), a two-seater, single-engine aircraft that could enter service in 2027/2028.

Completion of deliveries of the NH.90 Lobos to 48th Wing, 803rd Squadron, which will have as missions those of SAR/Combat SAR, MEDEVAC (Medical Evacuations), PR (Personal Recovery) and SAO (Special Air Evacuations). The programme includes a simulator manufactured by Indra and installed at Cuatro Vientos AB.

Under the heading of armaments, the purchase of 100 AiM-120C-7/7 Advanced Medium Range Air-to-Air Missiles (AMRAAM) is underway..

Shortly before the end of 2020, the EdA withdrew from service the twelve C-130 Hercules that were part of the 31st Wing for 42 years. Some of these aircraft will continue to fly far from Spain, as two have been sold to the Uruguayan Air Force, two to the Peruvian Air Force, and six will be used by Blue Aerospace, a company dedicated to providing spare parts. One of the two remaining aircraft will be handed over to the 'Museo del Aire' and the other will remain at the Zaragoza AB, in memory of an aircraft that for so many years accumulated 158,032 flight hours, transported 141,865,206 kilos of cargo, and carried almost one million passengers.

2
THE FORCE

AIR FORCE HEADQUARTERS

The Air Force Headquarters (Cuartel General del Ejército del Aire), located in the centre of the city of Madrid, was built between 1939 and 1953. It is situated in an area that during the Spanish Civil War (1936–1939) was a disputed territory by the contending sides and due to that reason, was destroyed.

Until 1977, it housed the so-called Ministry of Air, a state agency that administered everything related to aviation, both civil and military.

It was from the arrival of democracy in Spain that the separation of civil and military institutions took place, with the building then becoming the EdA's Headquarters.

From that time onwards, this large building has housed the following institutions:

Ejército del Aire Headquarters, Madrid. (Author's Photo)

AIR FORCE STAFF (ESTADO MAYOR DEL AIRE, EMA)

- Head of Technical Services, and Information and Telecommunications Systems (Jefatura de Servicios Técnicos y de Sistemas de Información y Telecomunicaciones, JSTCIS)
- EdA's Historical and Cultural Service (Servicio Histórico y Cultural del Ejército del Aire, SHYCEA)
- EdA's Headquarter Group (Agrupación del Cuartel General del Ejército del Aire, ACGEA)
- Cabinet of the EdA's Chief of Staff (Gabinete del Jefe del Estado Mayor del Ejército del Aire, GABJEMA)
- Air Legal Advisory (Asesoría Jurídica del Aire, AJA)
- Audit Office (Intervención Delegada Central, IDC)
- Economic Affairs Directorate (Dirección de Asuntos Económicos, DAE)

AIR FORCE CHIEF OF STAFF

In addition to those mentioned above, there are also the following institutions:

Air Force Staff (Estado Mayor del Aire, EMA)

Second Chief of EdA's Staff (Segundo Jefe del Estado Mayor del Ejército del Aire, SEJEMA)

Head Office:

- General Secretariat of Staff (Secretaría General del Estado Mayor)
- Command Post of the EdA's Chief of Staff (Puesto de Mando del Jefe del Estado Mayor del Ejército del Aire, PMJEMA)
- Plans Division (División de Planes, DPL)
- Operations Division (División de Operaciones, DOP)
- Logistics Division (División de Logística, DLO)

The EMA is the main institution of the JEMA (Jefe del Estado Mayor del Aire, Chief of the Air Staff), responsible for providing the necessary guidelines to support its decisions, turning them into orders and ensuring that they are complied with.

The Air Force Chief of Staff is an Air General, Lieutenant General (four stars) of the EdA General Corps.

GEOGRAPHIC LOCATION OF AIR BASES AND AERODROMES

All the Air Bases have a double purpose: on the one hand, they allow the deployment, instruction, training, and execution of the air actions and on the other hand, their supply and maintenance.

The Military Aerodrome fulfils the same double purpose, but with a restricted character as regards the operational capacity and the maintenance of the material of the EdA's Units.

The occasional aerodromes – also called deployment aerodromes – have only one guard detachment and minimum means to welcome punctually the aircraft in transit.

AIR BASES AND THEIR LOCATION	
Torrejón AB	Madrid
Getafe AB	Madrid
Cuatro Vientos AB	Madrid
Matacán AB	Salamanca
Villanubla AB	Valladolid
Morón AB	Sevilla
Talavera AB	Toledo
Armilla AB	Granada
San Javier AB	Murcia
Albacete AB	Albacete

Alcantarilla AB	Murcia
Zaragoza AB	Zaragoza
Son San Juan AB	Mallorca (Balearic Islands)
Gando AB	Gran Canaria (Canary Islands)
MILITARY AERODROMES AND THEIR LOCATION	
Barajas Military Aerodrome	Madrid
León Military Aerodrome	León
Pollensa Military Aerodrome	Mallorca (Balearic Islands)
Lanzarote Military Aerodrome	Lanzarote (Canary Islands)
Los Rodeos Military Aerodrome	Tenerife (Canary Islands)
Santiago Military Aerodrome	La Coruña
EVENTUAL AERODROMES AND THEIR LOCATION	
Palma del Río Aerodrome	Córdoba
Ablitas Aerodrome	Navarra
Griñón Aerodrome	Madrid
OTHER AERODROMES AND THEIR LOCATION	
Lanzarote Aerodrome, Air Deployment Base	Lanzarote (Canary Islands)

WINGS, SQUADRONS, GROUPS, AND THEIR LOCATION

A Wing is a higher entity unit under the command of a Colonel, and usually occupies the entire physical space of an air base.

A fraction of a Wing is called a Squadron. A Wing can have several Squadrons according to the availability of aircraft. It is usually commanded by a Commander.

There are independent Squadrons.

A generally independent unit, whose specific mission or place of deployment so advises, is called a Group. A Group is usually commanded by a Colonel or Lieutenant Colonel.

Sometimes organic and command changes cause the name of any of the units mentioned to be changed, a Squadron to become a Group, and vice versa, always in accordance with the needs of the moment.

ALAS (WINGS)	
Ala 11 de Fuerzas Aéreas	Morón AB
Ala 12 de Fuerzas Aéreas	Torrejón AB
Ala 14 de Fuerzas Aéreas	Albacete AB
Ala 15 de Fuerzas Aéreas	Zaragoza AB
Ala 23 de Fuerzas Aéreas	Talavera AB
Ala 31 de Fuerzas Aéreas	Zaragoza AB
Ala 35 de Fuerzas Aéreas	Getafe AB
Ala 37 de Fuerzas Aéreas	Villanubla AB
Ala Mixta 46 de Fuerzas Aéreas	Gando AB
Ala 48 de Fuerzas Aéreas	Cuatro Vientos AB
Ala 49 de Fuerzas Aéreas	Son San Juan AB
Ala 78 de Fuerzas Aéreas	Armilla AB

ESCUADRONES (Squadrons)		
111 Escuadrón 112 Escuadrón	Morón AB	Grupo 11, Ala 11
121 Escuadrón 122 Escuadrón	Torrejón AB	Ala 12
141 Escuadrón 142 Escuadrón	Albacete AB	Ala 14
151 Escuadrón 152 Escuadrón 153 Escuadrón	Zaragoza AB	Ala 15
221 Escuadrón	Morón AB	Grupo 22, Ala 11
231 Escuadrón 233 Escuadrón	Talavera AB	Ala 23

Albacete AB Workshops, CASA C.212 Aviocar, T.12-48, 352-48, from 35th Wing. (Author's Photo)

Getafe AB Workshops, Beechcraft F-33A Bonanza, E.24A-15, 421-07, of the 42nd Group. (Author's Photo)

311 Escuadrón 312 Escuadrón 313 Escuadrón	Zaragoza AB	Ala 31
351 Escuadrón 352 Escuadrón 353 Escuadrón	Getafe AB	Ala 35
371 Escuadrón 372 Escuadrón	Villanubla AB	Ala 37
402 Escuadrón 403 Escuadrón 409 Escuadrón	Cuatro Vientos AB	Ala 48, CECAF
422 Escuadrón	Villanubla AB	Grupo 42
431 Escuadrón 432 Escuadrón	Torrejón AB	Grupo 43
451 Escuadrón	Torrejón AB	Grupo 45
462 Escuadrón	Gando AB	Ala Mixta 46
471 Escuadrón 472 Escuadrón	Torrejón AB	Grupo Mixto 47
541 Escuadrón	Torrejón AB	Grupo 45
721 Escuadrón	Alcantarilla AB	Parachuting Military School
741 Escuadrón 744 Escuadrón	Matacán AB	Grupo Escuelas Matacán
781 Escuadrón 782 Escuadrón	Armilla AB	Ala 78
791 Escuadrón 792 Escuadrón 794 Escuadrón	San Javier AB	AGA
801 Escuadrón	Son San Juan AB	Ala 49 SAR
803 Escuadrón	Cuatro Vientos AB	Ala 48 SAR

Escuadrón de Apoyo Operativo a la Guerra Electrónica (ESAOGEL)	Torrejón AB
Escuadrón de Vigilancia Aérea No. 1 (EVA-1)	
Escuadrón de Vigilancia Aérea No. 2 (EVA-2)	
Escuadrón de Vigilancia Aérea No. 3 (EVA-3)	
Escuadrón de Vigilancia Aérea No. 4 (EVA-4)	
Escuadrón de Vigilancia Aérea No. 5 (EVA-5)	
Escuadrón de Vigilancia Aérea No. 7 (EVA-7)	
Escuadrón de Vigilancia Aérea No. 9 (EVA-9)	
Escuadrón de Vigilancia Aérea No. 10 (EVA-10)	
Escuadrón de Vigilancia Aérea No. 11 (EVA-11)	
Escuadrón de Vigilancia Aérea No. 12 (EVA-12)	
Escuadrón de Vigilancia Aérea No. 13 (EVA-13)	
Escuadrón de Vigilancia Aérea No. 21 (EVA-21)	
Escuadrón de Vigilancia Aérea No. 22 (EVA-22)	

GRUPOS (GROUPS)		
Grupo 11 de Fuerzas Aéreas Grupo 22 de Fuerzas Aéreas	Ala 11	Morón AB
Grupo 42 de Fuerzas Aéreas		Villanubla AB
Grupo 43 de Fuerzas Aéreas Grupo 45 de Fuerzas Aéreas		Torrejón AB
Grupo 82 de Fuerzas Aéreas		Gando AB
Grupo Mixto 47 de Fuerzas Aéreas		Torrejón AB
Grupo Escuelas de Matacán (GRUEMA)		Matacán AB
Grupo de Alerta y Control (GRUALERCON)		Gando AB
Grupo Norte de Mando y Control (GRUNOMAC)		Torrejón AB
Grupo Central de Mando y Control (GRUCEMAC)		Zaragoza AB
Escuadrilla de Apoyo al Despliegue Aéreo (EADA)		Zaragoza AB

Communications system ACAR La Muela (Zaragoza). (Author's Photo)

Zaragoza AB control tower. (Author's Photo)

Zaragoza AB, F-18B Hornet, C.15-22, 15-09. (Author's Photo)

BASES, UNITS, AND AIRCRAFT

TORREJÓN AIR BASE		
Ala 12	121 Escuadrón	F-18 A/B
	122 Escuadrón	
Grupo 43	431 Escuadrón	CL-215T
	432 Escuadrón	CL-415
Grupo 45	451 Escuadrón	A.310 Falcon 900
Grupo Mixto 47	471 Escuadrón	Falcon 20
	472 Escuadrón	C.212 ELINT
Escuadrón de Apoyo Operativo a la Guerra Electrónica (ESAOGEL)		
Centro Logístico de Armamento y Experimentación (CLAEX)	541 Escuadrón	It has aircraft temporarily assigned LANZA radar maintenance. Air weapons maintenance.
Mando Aéreo de Combate (MACOM)		
Grupo Central de Mando y Control (GRUCEMAC)		
Centro de Operaciones Aéreas Combinadas de la OTAN (CAOC-TJ)		
Mando de Defensa y Operaciones Aéreas (MDOA)		
Mando de Operaciones (MDOP)		
Centro de Operaciones de Vigilancia Espacial (COVE)		

GETAFE AIR BASE		
Ala 35	351 Escuadrón	CN-295
	352 Escuadrón	CN-295
	353 Escuadrón	CN-295

CUATRO VIENTOS AIR BASE		
Ala 48	402 Escuadrón	AS.332M-1 AS.532UL
	803 Escuadrón	AS.332B CN.235 VIGMA-SAR
Centro Fotográfico y Cartográfico (CECAF)	403 Escuadrón	Citation V, C.235
	409 Escuadrón	C.90
Maestranza Aérea de Madrid (MAESMA)		

MATACÁN AIR BASE		
Grupo Escuelas de Matacán (GRUEMA)	741 Escuadrón	C.101
	744 Escuadrón	C.235M
Escuela de Sistemas de Vuelo No Tripulado		

VILLANUBLA AIR BASE		
Ala 37	371 Escuadrón	C.212
	372 Escuadrón	
Grupo 42	422 Escuadrón	F-33

MORÓN AIR BASE			
Ala 11	Grupo 11	111 Escuadrón	Eurofighter
		112 Escuadrón	EF-2000
	Grupo 22	221 Escuadrón	P-3A/B
Segundo Escuadrón de Apoyo al Despliegue Aéreo (SEADA)			

TALAVERA AIR BASE		
Ala 23	231 Escuadrón	F-5B
	232 Escuadrón	B/MQ-9 Reaper

ARMILLA AIR BASE		
Ala 78	781 Escuadrón	S-76 EC.120B 'Patrulla Aspa'
	782 Escuadrón	

SAN JAVIER AIR BASE		
Academia General del Aire (AGA)	791 Escuadrón	T-35C
	792 Escuadrón	C.101
	794 Escuadrón	'Patrulla Águila'

ALBACETE AIR BASE		
Ala 14	141 Escuadrón	Eurofighter
	142 Escuadrón	
Maestranza Aérea de Albacete (MAESAL)		
Componente Nacional del Programa de Liderazgo Táctico (TLP)		

ALCANTARILLA AIR BASE		
Escuela Militar de Paracaidismo	721 Escuadrón	C.212
Escuadrón de Zapadores Paracaidistas (EZAPAC)		

ZARAGOZA AIR BASE		
Ala 15	151 Escuadrón	F-18A
	152 Escuadrón	F-18A
	153 Escuadrón	F-18B
Ala 31	311 Escuadrón	A.400
	312 Escuadrón	A.400
Grupo Norte de Mando y Control (GRUNOMAC)		
Escuadrilla de Apoyo al Despliegue Aéreo (EADA)		
Centro Nacional – European Tactical Airlift Centre (CN-ETAC)		

SON SAN JUAN AIR BASE		
Ala 49	801 Escuadrón	SA.330-H/J AS.332C-1 CN.235 VIGMA-SAR

GANDO AIR BASE
Mando Aéreo de Canarias (MACAN)

Alarm barracks at Torrejón AB. (Author's Photo)

A Mirage F.1EE in front of one of hardened aircraft shelters at the Albacete AB. (Author's Photo)

Ala Mixta 46	462 Escuadrón	F-18A
	Grupo 82, 802 Escuadrón	AS.332M-1 AS.332C-1 CN.235 VIGMA-SAR
Grupo de Alerta y Control (GRUALERCON)		

LEÓN MILITARY AERODROME
Academia Básica del Aire (ABA)

TABLADA AIR QUARTERING, ACAR TABLADA
Grupo Móvil de Control Aéreo (GRUMOCA)

OTHER INSTALLATIONS OF THE EJÉRCITO DEL AIRE

AIR QUARTERINGS
Acuartelamiento Aéreo de Getafe, ACAR Getafe (Madrid): CLOMA, CLOTRA, GRUAT, Escuela de Automovilismo, GLUTRA, ESTRAM-1 *
Acuartelamiento Aéreo Alto de los Leones, ACAR Alto de los Leones (Segovia): ESTRAM-3
Acuartelamiento Aéreo El Prat, ACAR El Prat (Barcelona)
Acuartelamiento Aéreo de Bobadilla, ACAR Bobadilla (Málaga)
Acuartelamiento Aéreo El Vedat, ACAR El Vedat (Valencia): ESTRAM-4
Acuartelamiento Aéreo La Muela, ACAR La Muela (Zaragoza): ESTRAM-6
Acuartelamiento Aéreo Las Palmas, ACAR Las Palmas de Gran Canaria
Grupo Cuartel General del Mando Aéreo de Canarias
Acuartelamiento Aéreo de Tablada, ACAR Tablada (Sevilla): Centro Deportivo, Centro de Farmacia, Centro de Mobilización-2, Dirección de Enseñanza (DEN), Escuadrilla de Circulación Aérea Operativa de Sevilla (ECAO-Sevilla/ECAO-2), ESTRAM-2, GRUMOCA, Maestranza Aérea de Sevilla
Acuartelamiento Aéreo El Frasno, ACAR El Frasno (Zaragoza): EVA-1
Acuartelamiento Aéreo Villatobas, ACAR Villatobas (Toledo): EVA-2

Acuartelamiento Aéreo Constantina (Sevilla), ACAR Constantina: EVA-3
Acuartelamiento Aéreo Rosas (Girona), ACAR Rosas: EVA-4
Acuartelamiento Aéreo Aitana (Alicante), ACAR Aitana: EVA-5
Acuartelamiento Aéreo Puig Mayor (Baleares), ACAR Puig Mayor: EVA-7
Acuartelamiento Aéreo Motril (Granada), ACAR Motril: EVA-9
Acuartelamiento Aéreo Barbanza (La Coruña), ACAR Barbanza: EVA-10
Acuartelamiento Aéreo Alcalá de los Gazules (Cádiz), ACAR Los Gazules: EVA-11
Acuartelamiento Aéreo Espinosa de los Monteros (Burgos), ACAR Espinosa de los Monteros: EVA-12
Acuartelamiento Aéreo Sierra Espuña (Murcia), ACAR Sierra Espuña: EVA-13
Acuartelamiento Aéreo Pozo de las Nieves (Las Palmas), ACAR Pozo de las Nieves: EVA-21
Acuartelamiento Aéreo Peña del Chache (Las Palmas), ACAR Las Palmas: EVA-22
* ESTRAM: Escuadrilla de Transmisiones, Transmission Squadron

OTHER SERVICES OF EJÉRCITO DEL AIRE
Sport Centres: eight
Colleges: two
Comandancia Militar Aérea de Melilla (CMAM–Melilla)
Escuadrilla de Circulación Aérea Operativa de Barcelona (ECAO-Barcelona/ECAO-4)
Jefatura de Movilidad Aérea (JMOVA), Zaragoza
Social Action Residence of Navacerrada (Madrid)

3
REGISTRATIONS, UNIT INDICATIVES AND OTHER ELEMENTS

Since the beginning of aviation's military activity, there has been a worldwide practice based on coding everything that flies with letters and numbers. Over the years and with the multiplication of aircraft, this coding has become so widespread that air forces have been obliged to establish codes by speciality, which makes it theoretically, easier to identify both the air units and the aircraft attached to them.

In the Spanish case, there are clear periods in which different systems have been used. Before the Civil War, aircraft were distinguished by groups of large capital letters on both sides of the fuselage and on the front and back of the wings. This system was later replaced by digits, which evolved and became more complicated during the war.

After the war, the new EdA adopted a method that simplified the system, and this has been passed down with some modifications.

NOMENCLATURE: MILITARY REGISTRATION OR SPANISH CODES

Consists of a set of letters, numbers, full stops, and hyphens, which define the speciality of the aircraft. The table shows the letters currently used by the aircraft whose basic mission is described:

A	Attack	Aircraft designed to attack enemy targets on ground or at sea
C	Fighter	Aircraft designed to intercept and destroy aircraft in the air
D	Rescue	Aircraft or helicopter specialised in search and rescue missions (SAR)
E	Trainer	Aircraft or helicopter whose main mission is flight training
H	Helicopter	All kinds of helicopters
P	Patrol	Highly autonomous multi-engine aircraft, suitable for maritime reconnaissance, anti-submarine, and minefield missions
N	–	Drone/Remote Operated Aircraft
T	Transport	Aircraft or helicopter intended for the transport of cargo and/or passengers
U	Utilitarian	Aircraft or helicopter adapted to carry out various missions such as observation, target dragging, small transport, liaison, firefighting, et cetera
X	Experimental	Aircraft dedicated to experimentation

Examples: T.19, transport aircraft CASA, CN.235.

The combination of these letters with each other, and with those in the following table, define the secondary mission of the aircraft that carry them. These are generally aircraft and/or helicopters permanently adapted to these missions. These second letters appear to the right of the one describing the main speciality.

D	Rescue	Modified aircraft or helicopter for SAR missions. Firefighting aircraft
E	Trainer	Aircraft or helicopter modified for flight training
K	Tanker	Multi-engine aircraft for transport or refuelling (including in-flight)
L	Large	Aircraft whose version is longer than the original model
M	Electronic	Aircraft modified and equipped to carry out the following missions: Electronic countermeasures Advanced radar Communications relay Calibrations and navigational aids Meteorology ELIN Missions
R	Reconnaissance	Modified aircraft for air reconnaissance missions. Aerial photography
T	Transport	Modified aircraft or helicopter for cargo or passenger transport

Examples: AE.9, CASA/Northrop F-5 assigned to the Fighter and Attack School. UD.13, Canadair CL-215 specialised in firefighting.

This combination of one or two letters will always be followed by a full stop, never a hyphen, which will separate them from one or more digits which particularise each type of aircraft (examples: E.26 corresponds to the ENAER T-35C Pillán or Tamiz; HE.24 corresponds to the Sikorsky S-76 Spirit school helicopter).

CASA C-101EB-1 Aviojet. Registration and Saint Andrew's Cross, E.25-27, 79-27. (Author's Photo)

Sometimes a letter appears after these digits, which can be A, B, C or D. These can be used to distinguish between different versions of the same aircraft or helicopter (e.g., T.12B).

After these two digits there is a hyphen, never a full stop, to separate the above from the digits that distinguish each of the aircraft (e.g.: T.12-2, which corresponds to the second Aviocar put into service). In some cases, these two digits are preceded by a zero (e.g.: AE.9-020, corresponding to an F-5).

Thus, we see that a military registration is constituted as follows: TK.23-02, which is the second A400M Atlas, whose main mission is the transport of cargo or passengers, and the secondary mission is in-flight refuelling.

The registrations assigned to the aircraft are painted in black on the tail, and although they tend to be small, they are always legible. As with cars, the registration numbers on military aircraft are assigned for life and are removed when the aircraft is taken out of service due to obsolescence or accident.

CURRENT REGISTRATIONS (YEAR 2022)

SPANISH CODE (REGISTRATION)	GENERIC NAME
CE.15/C.15	F-18 Hornet
C/CE.16	EF-2000
AE.9	F-5
TM.11	Mystère 20D/E/F Falcon 20
T.12B/C/D/TR.12A/T.12/TM.12	C.212 Aviocar, Series 100 and 200
T.21	C.295
T.18	Falcon 900
T.19A/T.19B	CN.235
TR.20	Citation V
E.22	C.90 King Air
E.24A	F-33C Bonanza
E.25	C.101 Aviojet
E.26	T-35 Pillán/Tamiz
UD.13T	CL-215/CL-215T/CL-415
HD.21/HT.21	AS.332 Super Puma/AS.532 Cougar
HD.21	AS 332C1 (H 215) Super Puma
HE.24	S-76C Spirit
HE.25	EC.120B Colibri
HT.27	AS.532UL Cougar
T.22	Airbus A.310-304
T.23	Airbus A.400 Atlas
T/TK.24	Airbus A.330-200 (MRTT)
D.4	CN.235 VIGMA-SAR
HD.29	Airbus Helicopters NH.90 Lobo
E.27	Pilatus PC-21
NR.5	General Atomics Predator B/MQ-9
–	Thales Group Fulmar
NR.03	Aerovironment Wasp AE RQ-12A
–	Aerovironment RQ-11 Raven

UNIT INDICATIVE

On the front or rear of the fuselage, depending on whether it is a fighter or a transport aircraft, there are two sets of digits separated by a hyphen or by the national roundel.

This numbering set distinguishes the unit to which the aircraft or helicopter belongs and the order number of the aircraft within the unit.

The Unit Indicative (formerly Unit Code), unlike the registration, changes whenever the unit is changed. This means that aircraft which, due to their characteristics or use, change location, will collect different Unit Indicatives during their operational life.

These codes can also be replaced because of administrative changes at the same air base, such as when a Squadron becomes a Wing or vice versa.

Let us take an example: F-18 Hornet, registration C.15-44, and Unit Indicative 12-02. The code C.15 always corresponds to the F-18, 44 being the number that corresponded to it when it was assigned the registration. This code remains unchanged throughout the operational life of the aircraft. The Unit Indicative 12-02 means that the aircraft belongs to 12th Wing of Torrejón and 02 is the number of the aircraft within the Wing.

This leads one to establish that the air units of the EdA are classified by means of a number that indicates their use, according to the following table:

NUMBER	OPERATIONAL EMPLOYMENT
1	Air Defence
2	Tactical (attack)
3	Transport
4	Mixed
7	Teaching
8	Search and Rescue

Currently there is no unit starting with 5 (those that would be available to the EdA's Staff), 6 (for training) and 9 (Command).

There are other elements that distinguish the units by themselves, such as the official emblems (the unofficial ones are usually short lived) which are sometimes shown on the tail of the aircraft, under the registration, or somewhere else on the fuselage. These emblems have evolved over time. Initially of large size, they have now become smaller and all that remains of the bright colours of yesteryear, is a mimetic sketch.

The inclusion of the Cross of Saint Andrew (X) on all EdA's aircraft is a custom inherited from the time of the Civil War. It distinguished the nationalist aircraft from the republican ones, and thus avoided shooting down those of their own side. Today, this Cross is much more discreet and has even been mimicked.

High volume aircraft, such as Atlas, display the number assigned to the aircraft inside the unit at two front points: on the radome and on the vertical landing gear cover. This is a way of distinguishing them when they approach from the front.

As far as aircraft nicknames are concerned, the solutions are very varied. For those coming from foreign manufacturers, normally their original nickname has been preserved (Hornet, Orion, et cetera) respecting the English spelling. However, on other occasions, this criterion has not been followed, as in the case of the F-5 which, although manufactured in Spain, could have inherited its original name of 'Freedom Fighter'. Perhaps it was a too long name? Instead, it has been familiarly nicknamed 'bicicleta'.

In this field, there are several intermediate cases, such as the nickname 'plancheta' (small iron) with which the crews affectionately baptised the Mirage III, although it was always known as Mirage III.

In the case of the Spanish-made aircraft in service today, CASA decided to hold a competition among the staff's children. The name 'Mirlo' was chosen, although it is designated as Aviojet or simply C.101. As for the C.212, from the beginning it was designated as Aviocar and the name has remained.

As for the CN.235, it officially has no nickname. It is simply known as the CN.235 or 235, although on many occasions even its crew members call it 'Nurtanio' – perhaps because of the exoticism, although this is an aberration since it is neither a Spanish name nor does it correspond to anything related to those manufactured in Spain.

The author is in favour of describing a type of aircraft in full, followed by its registration number. Although this may seem long and complicated, it is more accurate. For example, in the case of the F-18 it should be said: McDonnell Douglas F-18A Hornet, C.15.

SERIAL NUMBER

Another way of distinguishing aircraft – in this case, those from the USA – is by the Serial Number, the numerical coding assigned to them when their manufacture is contracted, both for that country's own armed forces and for export. This number and other indications are usually printed on the outside of the fuselage and on the side where the pilot enters the cockpit. In Spain, the custom of leaving them in place was maintained for many years, although today such numbering is not in sight and is not even quoted.

CONSTRUCTOR NUMBER

Like any other element produced by the industry, each aircraft carries what is usually called a 'patent', a metal plate on which the main manufacturer and product data have been printed or stamped. Here is found the factory number which is the same as the one on the documentation. This 'patent' is hardly visible as it is located inside the cabin.

In the case of the C.212 Aviocar, CASA is in the habit of painting it on the tail, so it is perfectly visible. Aerospatiale, the manufacturer of the Puma, Super Puma and Cougar, also includes this number on the tail.

SPECIAL MARKINGS ON SOME EJÉRCITO DEL AIRE AIRCRAFT

For some decades, the EdA adopted, for aircraft carrying out representative missions, an indication related to their origin and mission.

Thus, the old Boeing B-707, and the Mystère 20 and Falcon 900, of 45th Group and later 47th Group, were labelled 'Fuerza Aérea Española'.

In 2016, the regulations were changed and the label 'Reino de España' was adopted. Thus, it can be seen that the Airbus 310s and the five Falcon 900s, as well as the Bombardier CL-415s, have also adopted this form of distinction. The A.310s and Falcon 900s carry it on the top of the fuselage, whilst the CL-415s carries it above the Unit Indicative.

A.10C, Unit Indicative 214-114. (Author's Photo)

Unit Indicative: 793-27, C-101AB Aviojet, E.25-27, 79-27. (Author's Photo)

The A.310s and Falcon 900s also bear the inscription 'Ejército del Aire' in small size and on the tail.

AIRCRAFT IN SERVICE (OCTOBER 2022)

McDonnell Douglas F-18A and F-18B Hornet (C.15 and CE.15)

The FACA Programme for the selection of the new fighter aircraft began in 1978 and was to form the backbone of Spain's air potential until the end of the twentieth century.

At the time, no doubt influenced by NATO's recent decision in favour of the General Dynamics F-16, this fighter appeared to be the most qualified candidate to replace the old Spanish Phantom, even if, in the following months, new options were considered with all the US fighters in production or testing being considered, from the simplest of all, the Northrop F-5E Tiger II, to the expensive Grumman F-14A Tomcat and McDonnell-Douglas F-15A Eagle, via the F-18A and General-Dynamics F-16A Fighting Falcon.

As time went by, the Tiger II, which did not fully meet the specifications of the EdA, and the F-14 and F-15, due to their high cost, were discarded, leaving only the F-16 and F-18A.

At the 1979 Le Bourget Air Show, the Dassault Mirage 2000 fighter was presented as a European alternative. The complexity of the selection process, the incessant increase in costs, political pressures and the modifications that manufacturers made to their models, delayed the final decision and reduced the number of aircraft to be purchased. At the beginning of the 1980s, the Tornado was also chosen. It was in July 1982 that the Spanish government decided to purchase eighty-four F-18A.

A few months later, following a change of government, the FACA programme was completely reconsidered and the Tornado was chosen as a fully European option. In May 1983, the decision was finally taken to purchase the F-18A, although the number of examples was reduced to 72.

The decision to acquire the F-18A Hornet revealed the EdA's desire to provide Air Combat Command with a twin-engine aircraft with a high strike and penetration capability. Perhaps lacking the degree of specialisation offered by the separate intercept or strike versions of the Tornado, the Hornet was an outstanding weapon system capable of broadly matching the Tornado's capabilities, which it was to surpass in air combat. Over the years, the Hornet has evolved into a combat aircraft of great potential.

Initiated in the late 1960s by Northrop as the P-600 Cobra, the design conceived as a continuation of the F-5 philosophy, gradually evolved into the YF-17 prototype, which flew for the first time on 9 June 1974. In the so-called 'Contract of the Century', it also competed against the General Dynamics F-16, the Dassault Mirage F-1E and the Swedish SAAB Viggen.

Shortly thereafter, the US Navy became interested in it as a future carrier-borne aircraft to replace its Phantom and Corsair II. This possibility meant that the YF-17 would have to move from being a light fighter to become the high-potential F-18A fighter-bomber. This also meant that Northrop, the parent aerospace firm, would have to take on the role of prime subcontractor, leaving McDonnell Douglas, with its extensive experience in naval aircraft, to lead the programme.

Although subject to some initial criticism, both in its home country and in Europe because of problems typical of the early stages of development, the Hornet is undoubtedly a great weapon system. Not only have the US Navy and the US Marine Corps added it to their inventory, but other countries in the western area, such as

Serial Number 57-645, Lockheed T-33A-10-LO. (Author's Photo)

Serial Number 51-13417, North American F-86F-25-NA Sabre, C.6-235. (Author's Photo)

Australia, Canada, and Switzerland, also believed in its capability. The first EdA's F-18As registered C.15 arrived at the Zaragoza base in July 1986 for incorporation into the 151st Squadron of the 15th Wing that had been created for them. Subsequently, other aircraft joined the 12th Wing at Torrejón de Ardoz to replace the F-4C Phantom. The F-18 Hornet allowed the EdA to carry out a complete programme for the first time, from the evaluation of the aircraft to the declaration of 'ready for combat'.

The seventy-two Hornets purchased from McDonnell-Douglas (now the Boeing Company) were delivered between 1986 and 1990. In September 1995, twenty-five Hornets of the same type were

Registration and constructor number: T.12B-66, C-212AB-1-5-129. (Author's Photo)

purchased from the US Navy under the CX Programme to cover aircraft retired from service. The first aircraft were introduced, after major overhaul and modernisation, between 1996 and 2000.

REGISTRATION	UNIT INDICATIVE	SERIAL NUMBER	CONSTRUCTOR NUMBER	CRASHED
CE.15-1	15-70	162846	840371	
CE.15-2	15-71	162852	840381	
CE.15-3	15-72	162860	840395	
CE.15-4	15-73	162873.	840415	
CE.15-5	15-74	162883	840431	
CE.15-6	15-75	162887	840438	
CE.15-7	15-76	162896	840451	
CE.15-8	12-71	162902	840461	
CE.15-9	15-77	163108	840500	
CE.15-10	12-73	163121	850523	
CE.15-11	12-74	163131	850541	
CE.15-12	12-75	163141	850558	
C.15-13	12-01	162862	840399	
C.15-14	15-01	163100	850487	
C.15-15	15-02	163144	850510	
C.15-16	15-03	163118	850517	
C.15-17	15-04	163124	850529	15.08.1994
C.15-18	15-05	163126	850533	
C.15-19	15-06	163133	850545	07.11.1988
C.15-20	15-07	163139	850554	
C.15-21	15-08	163145	850565	
C.15-22	15-09	163149	850572	
C.15-23	15-10	163153	850579	
C.15-24	15-11	163155	850583	
C.15-25	15-12	163160	850591	
C.15-26	15-13	163163	850597	
C.15-27	15-14	163156	850601	
C.15-28	15-15	163169	850608	
C.15-29	15-16	163170	850610	
C.15-30	15-17	163173	850616	
C.15-31	15-18	163430	850625	
C.15-32	15-19	163431	850627	
C.15-33	15-20	163432	850629	
C.15-34	15-21	163437	850638	
C.15-35	15-22	163439	850642	
C.15-36	15-23	163446	850653	
C.15-37	15-24	163448	850656	
C.15-38	15-25	163453	850667	
C.15-39	15-26	163445	850670	
C.15-40	15-27	163459	850677	
C.15-41	15-28	163461	850681	
C.15-42	15-29	163471	850697	13.03.2000
C.15-43	15-30	163473	850700	
C.15-44	12-02	163485	860718	
C.15-45	12-03	163486	860720	
C.15-46	12-04	163499	860740	
C.15-47	15-31	163500	860743	
C.15-48	12-06	163702	870762	
C.15-49	12-07	163703	870765	
C.15-50	12-08	163716	870780	
C.15-51	12-09	163717	870781	
C.15-52	12-10	163718	870785	17.10.2017
C.15-53	12-11	163730	870799	
C.15-54	12-12	163731	870801	
C.15-55	12-13	163745	870818	
C.15-56	12-14	163746	870820	
C.15-57	12-15	163754	870830	
C.15-58	12-16	163756	870833	02.04.2009
C.15-59	12-17	163761	870840	
C.15-60	12-18	163762	870842	

McDonnell Douglas, F-18B Hornet, CE.15-2, 151-02 (15-71). (Author's Photo)

McDonnell-Douglas F-18 Hornet, C.15-33, 15-20. (Author's Photo)

C.15-61	12-19	163773	870854	
C.15-62	12-20	163775	870857	
C.15-63	15-32	163779	870862	13.03.2000
C.15-64	15-34	163781	870865	
C.15-65	12-23	164008	880892	
C.15-66	12-24	164010	880894	
C.15-67	15-33	164025	880910	
C.15-68	12-26	164426	880912	
C.15-69	12-27	164038	880925	
C.15-70	12-28	164039	880927	
C.15-71	12-29	164050	880939	28.11.1991
C.15-72	12-30	164051	880941	
C.15-73	46-01	161936	140	
C.15-74	46-02	162415	251	16.06.2009
C.15-75	46-03	162416	252	
C.15-76	21-04	162426	266	21.07.1999
C.15-77	46-05	162446	294	
C.15-78	46-06	162471	328	16.06.2009
C.15-79	46-07	161940	146	
C.15-80	46-08	161944	151	
C.15-81	46-09	161949	158	
C.15-82	46-10	162456	308	
C.15-83	46-11	162461	314	
C.15-84	46-12	162465	320	
C.15-85	46-13	161951	161	
C.15-86	46-14	161939	144	
C.15-87	46-15	161950	159	
C.15-88	46-16	161953	163	
C.15-89	46-17	162421	258	
C.15-90	46-18	162474	332	
C.15-91	46-19	161926	127	11.02.2003
C.15-92	46-20	161935	139	
C.15-93	46-21	161954	165	
C.15-94	46-22	161958	170	
C.15-95	46-23	161977	197	
C.15-96	46-24	162444	291	

Eurofighter Typhoon (XCE.16, CE.16, C.16)

Eurofighter was one of the first consortia made up entirely of European countries, specifically the United Kingdom, Germany, Italy, and Spain, to initiate a cooperative programme for the development of a state-of-the-art fighter aircraft, with the common goal of closing the technology gap between Europe and the United States. The creation of the first demonstrator for the future EFA designated as EAP (Experimental Aircraft Programme) represented a fundamental step forward for the Eurofighter Programme. It integrated such innovative solutions as the massive use of composite materials such as carbon fibre and aluminium-lithium alloys, as well as the development of display and monitoring systems for practically all the aircraft's parameters through multifunction screens. The result being a highly manoeuvrable, twin-engine, multi-role fighter designed and produced by the consortium of European companies: AIRBUS, BAE SYSTEMS, EADS CASA and Alenia Aeronáutica.

The first flight of the DA01 prototype took place on 27 March 1994. During this development phase, a total of seven aircraft were built, each with very specific tasks. These development aircraft were followed by seven more called IPA (Instrumented Production Aircraft), standard production aircraft prepared for the development of future systems to be integrated into series production aircraft. In the case of Spain, these were the IPA4 and IPA9, based at the Airbus Getafe facilities, to which was added the ISPA (Instrumented Series Production Aircraft) formally assigned to 11th Wing and is regularly flown by CLAEX personnel.

In 1998, the production agreements between the four participating countries were for 620 aircraft built on the basis of production contracts that would involve three blocks and a progressive increase in capacity. This planning was surprised both by the economic crisis and the progressive indebtedness of the countries participating in the programme, also by the delay in the development of certain key equipment, as well as the lack, at that time, of customers for the export of the aircraft. This led to the postponement of the development of block 3, the most advanced version of Eurofighter, which was finally undertaken in two phases in 2009.

Continued on page 24

At the EdA, on 31 August 1996, the two-seater prototype DA06 made its first flight at Getafe. Assembled by EADS CASA, the aircraft was registered XCE.16-01.

The initial contract was for 87 aircraft which, for budgetary reasons, was reduced to 73 aircraft, of which 14 were two-seater and 58 single-seater and one instrumented (this was returned to Germany).

Negotiations began in mid-2020 for 20 new Typhoons, which will be delivered between 2026 and 2030.

The contract was signed in June 2022. The 20 aircraft contracted are 16 single-seat and four twin-seat aircraft, plus an advanced simulator, 48 engines and eight spare engines. As mentioned, deliveries will take place between 2026 and 2030. Although these aircraft are initially intended to replace the F-18 Hornets of the 46th Wing at Gando, it is possible that they will also equip Wings 11th or 14th.

During its operational life and until summer 2022, a total of four aircraft have been lost in accidents (see list).

In December 2021 they were distributed as follows:

13 two-seaters (all in-flight, plus one lost = 14)
- 11 in 11th Wing of Morón AB
- 2 in 14th Wing of Albacete

56 single-seater (in-flight in February 2021, plus three lost = 59):
- 29 in 11th Wing at Morón AB (of a total of 31, two had accidents)
- 27 in 14th Wing of Albacete AB (of a total of 28, one of them had an accident)

TYPE	REGISTRATION	UNIT INDICATIVE	CONSTRUCTOR NUMBER	WITHDRAWN	CRASHED
	XCE.16-01		DA6		21.11.2002
*	XC.16-02		DA1	March 2004	
14 TWO-SEATER TRAINING AIRCRAFT:					
	CE.16-01	11-01	ST001 †		
	CE.16-02	11-02	ST002		
	CE.16-03	11-03	ST003		
	CE.16-04	11-04	ST004		
	CE.16-05	11-05	ST005		
	CE.16-06	11-06	ST006		
	CE.16-07	11-07	ST007		
	CE.16-08	11-77	ST008		24.08.2010
	CE.16-09	11-09	ST–		
	CE.16-10	11-10	ST–		
	CE.16-11	14-70	ST–		
	CE.16-12	14-71	ST10000		
	CE.16-13	11-13	ST–		
	CE.16-14	11-14	ST10015		
59 SERIAL AIRCRAFT:					
	C.16-20	11-91	IPA4/PS001 †		
	C.16-21	11-21	SS001 †		
	C.16-22	11-22	SS002		
	C.16-23	11-23	SS003		
	C.16-24	11-24	SS004		
	C.16-25	11-25	SS005		
	C.16-26	11-26	SS–		
	C.16-27	11-27	SS–		
	C.16-28	11-28	SS–		
	C.16-29	11-29	SS–		
	C.16-30	11-30	SS011		
	C.16-31	14-31	SS–		
	C.16-32	11-32	SS–		
	C.16-33	11-33	SS–		
	C.16-34	11-14	SS015		09.06.2014
	C.16-35	11-35	SS–		
	C.16-36	14-03	SS–		
	C.16-37	14-04	SS–		
	C.16-38	11-38	SS–		
	C.16-39	14-06	SS021		
	C.16-40	11-40	SS–		
	C.16-41	11-41	SS021		
	C.16-42	11-42	SS023		
	C.16-43	11-43	SS–		
	C.16-44	14-09	SS–		
	C.16-45	14-10	SS–		
	C.16-46	11-46	SS–		
	C.16-47	14-11	SS–		
	C.16-48	14-12	SS–		
	C.16-49	14-13	SS030		
	C.16-50	11-30	SS–		
	C.16-51	11-51	SS–		
	C.16-52-10001	11-52	SS–		
	C.16-53-10002	11-53	SS034		
	C.16-54-10003	11-54	SS–		
	C.16-55	11-55	SS–		
	C.16-56-10007	11-56	SS–		
	C.16-57-10012	11-57	SS10012		
	C.16-58	14-	SS–		
	C.16-59	14-	SS–		
	C.16-60	14-	SS–		
	C.16-61	14-	SS–		
	C.16-62	14-	SS–		
	C.16-63	11-63	SS–		
	C.16-64	14-	SS–		
	C.16-65-10054	14-23	SS–		
	C.16-66	14-	SS–		
	C.16-67	14-	SS–		
	C.16-68-10091	14-	SS–		
	C.16-69-10092	14-	SS10092		12.10.2017
	C.16-70-10145	14-28	SS051		2017
	C.16-71	14-	SS–		
	C.16-72	14-	SS–		
	C.16-73	14-	SS–		
	C.16-74	14-32	SS–		
	C.16-75	14-	SS–		
	C.16-76	14-	SS–		
	C.16-77-10234	14-35	SS10234		

	C.16-78-10235	14-36	SS10235		
* XC.16-02 was temporarily loaned to EADS CASA by Germany and returned after a time. † ST: two-seater. IPA: Instrumented Production Aircraft. SS: single-seater.					

CASA-Northrop F-5A, RF-5A and F-5B/M (AE.9, A.9 and AR.9)

At the end of 1964, the Spanish Government made public the decision to equip the EdA with a series of seventy F-5s, in versions A and B, manufactured under licence at the Construcciones Aeronáuticas, S.A. factories.

From that moment on, CASA practically focused on the Azor Programme and some minor participation in the German HFB-320, began a plan to modernise its facilities and machinery that would achieve practical results in the following decade. For 4 years, CASA had to absorb the latest aeronautical technology and prepare its facilities and technicians for a programme costing a total of 2.8 billion pesetas at the time (16,830 €) and of unprecedented importance in its history. At the agreed time, Northrop sent eight F-5Bs in various stages of manufacture to Spain to familiarise CASA operators with the new aircraft.

Specifically, the first two aircraft were in the final assembly phase, the next three were awaiting instrumentation, and the last three were fully disassembled. The engines and equipment were ready to be installed in the aircraft. On 22 May 1968, the first of the F-5Bs assembled at CASA's Getafe plant made its first flight at the Getafe Air Base. A few months later and after many tests, this aircraft, registered CE.9-001, was delivered to the EdA together with 10 other aircraft of the same type.

In early January 1970, the first F-5Bs joined 202nd Squadron based at Morón, in order to familiarise the pilots of the unit before the arrival of the F-5A single-seaters that were destined for them. When these F-5As arrived in September of that year, 204th Squadron was created. These two units were short lived, as with the arrival of all the F-5Bs, they were disbanded and a single squadron, 211th Squadron, was formed, based at Talavera, the headquarters of the Jet School. These new aircraft replaced the veteran T-33As. To meet the need for aerial reconnaissance aircraft, the EdA's Command arranged for CASA to convert the last eighteen F-5A into RF-5A by installing KS-92 cameras in the nose. These 18 aircraft were given the registration numbers AR.9-053 to AR.9-070 and were assigned to the newly created 212th Squadron. By the end of 1971, all Spanish F-5s were in-flight. Undoubtedly, this EdA's fighter was the aircraft that achieved the highest cost/availability ratio. Without attack electronics, with two reliable engines and with extraordinary flight characteristics, the F-5 became a highly appreciated aircraft by Spanish pilots.

In the mid-1960s, the growing tension in North Africa as the decolonisation of the Spanish Sahara drew near, meant that the F-5s based at Morón were frequently deployed to the Canary Islands. Soon after and using the F-5A and RF-5A fleet, two mixed squadrons were created with attack and reconnaissance aircraft, which were assigned to the Gando Air Base in the Canary Islands. The arrival of Mirage F-1Es at Gando in 1982, brought the F-5s of 464th Squadron (which was disbanded) back to the Morón Base (212th Squadron).

In 2022, of those 70 aircraft, only 19 remain in service and have been upgraded to F-5B/M (see table below).

Airbus A.330-200 (MRTT)

In order to enhance long range transport and refuelling missions, the EdA acquired three of its A.330-200s from Iberia to convert them into transport aircraft, which, in the mid-2020s, became MRTT aircraft. Their original civilian registration numbers were, respectively, EC-MIL, EC-MSA and EC-MKI. The first of these was incorporated on 30 November 2021.

Registration	Unit Indicative	Constructor Number
TK.24-01	452-01	1694
TK.24-02	452-02	1700
TK.24-03	452-03	1719

AIRBUS DEFENCE AND SPACE A.400 ATLAS (T.23, TK.23)

The A.400 is a military transport aircraft designed to cover both strategic and tactical missions.

The aircraft is powered by four turboprop engines (TP400) with a maximum take-off weight of 141 tonnes, capable of carrying a maximum payload of 32 tonnes at 1,700 nautical miles at a maximum speed of 0.68 Mach. The minimum payload range is 4,400 nautical miles.

The aim of the programme was to obtain 170 units of the aircraft for the seven participating states, capable of replacing the ageing European fleets of Lockheed C.130 Hercules and C.160 Transall. In principle, the total number of aircraft committed for each nation was as follows: Germany 53 aircraft, France 50, Spain 27, United Kingdom 22, Turkey 10, Belgium seven and Luxembourg one. The programme considered the possibility of manufacturing a large number of aircraft for export. OCCAR (Organización Conjunta de Cooperación en Materia de Armamento, Organisation for Joint Armament Cooperation) was mandated by the seven participating states to manage the A 400M programme on their behalf and to act as the contracting authority. From the outset, an A 400M Programme Steering Board, composed of representatives of all participating nations, existed to oversee the programme at the strategic level and to issue high-level guidelines.

Five flight prototypes were built and flew more than 6,100 flight hours between December 2011 and December 2015. A further aircraft was destined for fatigue testing. By July 2014, four production aircraft had been delivered: the serial numbers were MSN007, MSN008 and MSN0010 to France, whilst MSN009 was delivered to Turkey. Two aircraft were then delivered to the United Kingdom and Germany, respectively. These deliveries made it possible to verify the performance and reliability of the A 400M model. The A 400M Memorandum of Understanding signed by the seven states in 2003, included the development and production of 170 aircraft at a fixed unit price in a single contract, known as the 'Design and Production Phase Contract'. That document also included options for initial logistics support, to be selected by each nation well in advance.

As mentioned above, the EdA was involved from the beginning in the development of the A.400M Atlas, deciding at the time, to contract 27 aircraft, but later announced that it would only incorporate 14, all of them destined for the 31st Wing based in Zaragoza.

The first aircraft delivered entered service at the end of 2016. A second one was delivered the following year. This, and one other aircraft have the capability to refuel fighters in the air.

The last aircraft were due for delivery in December 2022.

CASA-Northrop F-5A, A.9-046, 464-46 (21-08). (Author's Photo)

TYPE	REGISTRATION	UNIT INDICATIVE	CONSTRUCTOR NUMBER	WITH-DRAWN	CRASHED
F-5B/M	AE.9-001	23-23	2001		
F-5B	CE.9-002		2002		04.04.1974
F-5B	AE.9-003		2003		22.01.2003
F-5B	CE.9-004		2004		26.07.1972
F-5B/M	AE.9-005	23-25	2005		
F-5B	CE.9-006		2006		08.04.1973
F-5B/M	AE.9-007	23-01	2007		
F-5B/M	AE.9-008	23-02	2008		
F-5B/M	AE.9-009	23-03	2009		
F-5B/M	AE.9-010	23-04	2010		
F-5B/M	AE.9-011	23-05	2011		
F-5B/M	AE.9-012	23-06	2012		
F-5B	CE.9-013		2013		03.07.1974
F-5B/M	AE.9-014	23-07	2014		
F-5B	AE.9-015		2015		07.10.1991
F-5B/M	AE.9-016	23-09	2016		
F-5B/M	AE.9-017	23-10	2017		
F-5B/M	AE.9-018	23-26	2018		
F-5B	CE.9-019		2019		19.12.1972
F-5B	AE.9-020		2020		11.04.1989
F-5B	CE.9-021		2021		07.05.1975
F-5B/M	AE.9-022	23-12	2022		
F-5B	AE.9-023		2023		27.01.2006
F-5B/M	AE.9-024	23-14	2024		
F-5B	AE.9-025		2025		02.11.2012
F-5B	AE.9-026		2026		11.02.1982
F-5B/M	AE.9-027	23-16	2027		

REGISTRATION	FACTORY CODE	UNIT INDICATIVE	CONSTRUCTOR NUMBER
T.23-01	MSN44	31-21	0044
TK.23-02	10075	31-22	0070
TK.23-03	10076	31-23	0076
T.23-04	10174	31-24	0082
T.23-05	MSN97	31-25	
T.23-06		31-26	0098
TK.23-07	10208	31-27	0099
T.23-08	10217	31-28	0101
TK.23-09		31-29	
T.23-10	MSN111	31-30	
T.23-11	10220	31-31	112
T.23-12	10221	31-32	140
T.23-13		31-33	
T.23-14		31-34	

CASA C.212 Aviocar, Series 100 and 200 (XT.12, TR.12A, TR.12B, T.12B, T.12C, TE.12B and TM.12D)

The first studies of the Aviocar project date back to 1967, when CASA decided to undertake the design of a twin-engine cargo aircraft with small tonnage and STOL characteristics that could replace the old DC-3 and C.352 in service with the EdA. After a few years of research and development, the first prototype of the C.212, named XT.12-1 by military aviation, made its first flight, lasting 47 minutes, on 26 March 1971. In June 1971, this Aviocar was presented at the Le Bourget Air Show, whilst four months later, on 23 October, the second prototype, the XT.12-2, flew at Getafe. Again in 1973, with the test programme well advanced, this aircraft was again presented at Le Bourget, this time with four-bladed propellers, which would become the standard for the series to be produced. At the end of 1973, Portugal ordered 24 units,

F-5B/M	AE.9-028	23-17	2028		
F-5B/M	AE.9-029	23-18	2029		
F-5B/M	AE.9-030	23-19	2030		
F-5B/M	AE.9-031	23-20	2031		
F-5B	CE.9-032		2032		25.05.1976
F-5B	AE.9-033		2033		28.04.1989
F-5B	AE.9-034		2034		01.09.1988
F-5A	A.9-035		1035		07.10.1985
F-5A	C.9-036		1036		22.01.1977
F-5A	A.9-037		1037		02.12.1983
F-5A	A.9-038		1038	1993	
F-5A	A.9-039		1039	1993	
F-5A	A.9-040		1040	1993	
F-5A	A.9-041		1041	1993	
F-5A	A.9-042		1042	1993	
F-5A	A.9-043		1043		27.07.1979
F-5A	A.9-044		1044	1993	
F-5A	A.9-045		1045	1993	
F-5A	A.9-046		1046	1993	
F-5A	A.9-047		1047		08.09.1982
F-5A	A.9-048		1048	1993	
F-5A	A.9-049		1049		17.02.1984
F-5A	A.9-050		1050	1993	
F-5A	A.9-051		1051	1993	
F-5A	A.9-052		1052	1993	
RF-5A	AR.9-053		1053	1993	
RF-5A	AR.9-054		1054		25.09.1980
RF-5A	AR.9-055		1055	1993	
RF-5A	AR.9-056		1056	1993	
RF-5A	AR.9-057		1057		10.07.1981
RF-5A	AR.9-058		1058	1993	
RF-5A	AR.9-059		1059		12.01.1989
RF-5A	AR.9-060		1060	1993	
RF-5A	AR.9-061		1061		30.03.1998
RF-5A	AR.9-062		1062	1993	
RF-5A	AR.9-063		1063		18.05.1978
RF-5A	AR.9-064		1064	1993	
RF-5A	AR.9-065		1065		28.10.1986
RF-5A	AR.9-066		1066	1993	
RF-5A	AR.9-067		1067		10.07.1981
RF-5A	AR.9-068		1068	1993	
RF-5A	AR.9-069		1069	1993	
RF-5A	AR.9-070		1070		30.04.2003

the start of a very long order book, which included a significant number of examples built under licence in Indonesia.

Apart from the two prototypes mentioned (XT.12-1 and XT.12-2), which were destined for INTA, in 1971, the EdA purchased eight pre-series examples, six destined to carry out photographic reconnaissance missions (TR.12A) being equipped with Wild RC-10 cameras and destined for a unit based at Cuatro Vientos. The remaining two were dedicated to teaching at the San Javier Air Base and registered TE.12B-9 and TE.12B-10.

In 1973, thirty-two series aircraft (C.212-100, registered T.12B-11 to TE.12B-42) were contracted, whilst in 1976, a second series of nineteen C.212-100 were acquired, registered T.12B and T.12C. Later in 1977, a new series of ten C.212-100, registered T.12B, was contracted. Finally, in the 1980s, two more examples of the C.212-200 series were purchased, registered TM.12D-72 and TM.12D-73, equipped for ELINT (Electronic Intelligence) missions.

Much later, and from various sources in the civilian sector, the EdA incorporated eight aircraft (ex-Correos (Spanish mail service), ex-CASA and ex-Customs Surveillance Service).

Two of the T.12B version aircraft were equipped for SAR missions and therefore re-registered D.3A.

When they were decommissioned, some of the C.212 Aviocars were handed over to the air forces of various countries (Colombia and Bolivia), whilst others were installed as monument aircraft at various air bases, Spanish cities or places linked to Spanish aviation.

The C.212s served in the 35th Wing at Getafe, in the 461st Squadron of the 46th Joint Wing at Gando, in the 792nd Squadron at San Javier, in the 744th and 745th Squadrons of the Polimotor School of Matacán, in the 721st Squadron of the Paratrooper School of Alcantarilla and in the 403rd Reconnaissance Squadron at Cuatro Vientos.

By mid-2021, some 14 aircraft from 721st Squadron remain in service.

As of December 2021, the remaining C.212 Aviocars in service are in their final phase of operational life.

REGISTRATION	UNIT INDICATIVE	CONSTRUCTOR NUMBER	WITHDRAWN	CRASHED
TWO PROTOTYPES, C.212-10:				
XT.12-1		X-01	n/i †	
XT.12-2		X-02		12.03.1998
Eight pre-series aircraft, C-212-100:				
TR.12A-3		B-1-1	30.06.1999	
TR.12A-4		B-2-2	n/i	
TR.12A-5		B-3-3	n/i	
TR.12A-6		B-4-4	n/i	
TR.12A-7		B-5-5	07.12.1996	
TR.12A-8		B-6-6-08	n/i	
TE.12B-9		E-1-7	n/i	
TE.12B-10		E-2-8	n/i	
FIRST SERIES OF 32 AIRCRAFT, C.212-100:				
T.12B-11		A1-1-11		30.06.1976
T.12B-12		A1-6-16	n/i	
T.12B-13	72-13	A1-9-19		
T.12B-14		A1-2-12	2004	
T.12B-15		A1-5-15	20.11.2006	
T.12B-16		A1-10-20	n/i	
T.12B-17		A1-11-21	n/i	
T.12B-18		A1-12-22	March 2004	
T.12B-19		A1-13-23	n/i	
T.12B-20		A1-14-24	n/i	
T.12B-21		A1-15-27	n/i	
T.12B-22		A1-16-30	n/i	
T.12B-23		A1-17-31	2004	
T.12B-24		A1-18-40	30.06.2006	
T.12B-25		A1-19-43	n/i	
T.12B-26		A1-20-42	n/i	
T.12B-27		A1-21-46	n/i	
T.12B-28		A1-22-48	n/i	
T.12B-29		A1-23-51		09.10.2005
T.12B-30		A1-24-58	n/i	
T.12B-31		A1-25-59	December 2003	
T.12B-32		A1-26-65		02.01.1987
T.12B-33		A1-26-66	n/i	
T.12B-34		A1-28-67	n/i	
T.12B-35		A1-29-71	n/i	
T.12B-36		A1-30-72	n/i	
T.12B-37		A1-31-73	n/i	
T.12B-38		A1-32-74		06.02.1992
T.12B-39		A1-33-75	n/i	
TE.12B-40		E1-1-76	n/i	
TE.12B-41		E1-2-79	n/i	
TE.12B-42		E1-3-81		06.03.1991
SECOND SERIES OF 19 AIRCRAFT, C.212-100:				
T.12C-43		AV1-69	n/i	
T.12C-44		AV1-70	December 2003	
T.12B-45*		AA1-1-87	n/i	
T.12B-46		AA1-1-88	December 2003	
T.12B-47		AA1-1-94	n/i	
T.12B-48		AA1-4-99	December 2003	
T.12B-49	72-07	AA1-5-100		
T.12B-50		AA1-6-102	2002	
T.12B-51		AA1-7-104	2002	
T.12B-52		AA1-8-105	2004	

CASA C.212 T-12B-22, 352-22 (72-02). (Author's Photo):

T.12B-53		AA1-11-96	n/i	
T.12B-54		AA1-12-98	n/i	
T.12B-55	72-08	AA1-9-113	n/i	
T.12B-56		AA1-10-114	July 2003	
T.12B-57		AA1-13-110	2004	
T.12B-58		AA1-14-111	n/i	
T.12C-59		AV1-3-77	December 1973	
T.12C-60		AV1-4-109	2004	
T.12C-61		AV1-5-115	January 2010	
THIRD SERIES OF 10 AIRCRAFT, C.212-100:				
T.12B-62 **		AB1-1-121	n/i	
T.12B-63		AB1-2-122	n/i	
T.12B-64		AB1-3-123	January 2003	
T.12B-65	72-11	AB1-4-127		
T.12B-66	72-09	AB1-5-129		
T.12B-67	72-12	AB1-6-130		
T.12B-68		AB1-7-142	n/i	
T.12B-69	72-15	AB1-8-143		
T.12B-70	72-17	AB1-9-146		
T.12B-71	72-71	AB1-10-147		
TWO AIRCRAFT ELINT, C.212-200:				
TM.12D-72	47-12	DE1-1-313	n/i	
TM.12D-73		DE1-2-314		22.03.2000
EIGHT AIRCRAFT OF DIVERSE CIVIL ORIGIN, C.212-200:				
TM.12D-74 [1]		301 CC84	n/i	
TM.12D-75 [1]		270 CC84	2010	
TR.12D-76 [2]	72-21	359 VA25	n/i	
TR.12D-77 [2]	72-22	261 CE25	n/i	
TR.12D-78 [2]		311 FV25	2010	
TR.12D-79 [2]	72-23	323 CE25	n/i	
TR.12D-80 [2]		178 VF25	2010	
TR.12D-81 [2]	72-14		2010	

† n/i: the author has no information.
* T.12B-45 became D.3A-1 of the SAR. Upon being withdrawn, was delivered to Senegal.
** T.12B-62 became D.3A-2 of the SAR, where it was withdrawn on 02.04.14.
[1] TM.12D-74 and TM.12D-75 previously belonged to Spanish Postal Service (Correos) with the civil registrations EC-DTV and EC-DUC, respectively.
[2] TR.12D-76 to TR.12D-81 previously belonged to the Spanish Customs Surveillance Service (SVA), with civil registrations (respectively): EC-FAP, EC-FAQ, EC-DTL, EC-ECD, EC-DNB, EC-DRO. EC-FAP had been a demo aircraft owned by CASA, whilst EC-FAQ was purchased by the SVA in Argentina.

Marcel Dassault/Breguet Aviation Falcon 900 (T.18)

The 43rd Air Forces Group, based at Torrejón de Ardoz, has five Falcon 900 jets, a French-made executive tri-reactor.

The first two were received in April 1988 and three more were purchased from the Australian Air Force in June and July 2003.

This type of aircraft has a capacity to carry 14 passengers for flights of up to 7,000 kilometres non-stop.

TYPE	REGISTRATION	UNIT INDICATIVE	CONSTRUCTOR NUMBER
F 900A/B	T.18-1	45-40	038
F 900A/B	T.18-2	45-41	090
F 900B	T.18-3	45-42	077
F 900B	T.18-4	45-43	074
F 900B	T.18-5	45-44	073

CASA CN.235/10/100 VIGMA/SAR (T.19A and T.19B)

In 1982, whilst CASA was successfully producing its version of the Aviocar C.212-200, reinforced and with more powerful engines to increase its payload, an important and advantageous agreement was established with the Indonesian aeronautical firm Nurtanio, to develop a common project and produce a high-wing, multi-purpose transport capable of seating 38 people in a pressurised cabin. The CASA-Nurtanio CN.235 programme was scheduled to begin deliveries in 1985. The new aircraft was to be powered by two 1,700hp General Electric CT7-7 turboprops with a maximum speed of 483km/h and a maximum take-off weight of 13,000 kilos.

In Spain, the new development of CASA was suggested by the EdA, within the framework of the FATAM Programme, which wanted to strengthen its transport units with a medium-type aircraft equipped with a rear ramp.

In January 1989, the first copies of the Series 10 for the transport of VIP passengers arrived.

Between March 1992 and November 1993, the remaining 18 aircraft were delivered in a military version (100 series), for cargo and paratrooper drops.

From the early 2000s onwards, eight of these CN-235-100 began to be converted into specialised aircraft for maritime surveillance (VIGMA) and Air Rescue Service (SAR). For this mission, they were equipped with a Fully Integrated Tactical System (FITS), a maritime target search and tracking radar, FLIR/EO camera and AIS systems.

TYPE	REGISTRATION	UNIT INDICATIVE	CONSTRUCTOR NUMBER	WITHDRAWN
CN.235-10	T.19A-01 TR.19A-01	 403-01	13	
CN.235-10	T.19A-02 TR.19A-02	 403-2	14	
CN.235-100	T.19B-03		34	19.02.2011
CN.235-100	T.19B-04		35	19.02.2011
CN.235-100	T.19B-05 D.4-06	 SAR	36	July 2009
CN.235-100	T.19B-06 D.4-05	 SAR	37	April 2009
CN.235-100	T.19B-07	74-25	38	
CN.235-100	T.19B-08 D.4-04	 SAR	39	February 2009
CN.235-100	T.19B-09 D.4-02	 SAR	40	2008
CN.235-100	T.19B-10 D.4-03	 SAR	46	2008

Continued on page 30

CN-235-100	T.19B-11 D.4-07	SAR	47	2012
CN.235-100	T.19B-12 D.4-01		50	February 2008
CN.235-100	T.19B-13	74-31	54	
CN.235-100	T.19B-14 D.4-08	SAR	59	2012
CN.235-100	T.19B-15	74-33	60	
CN.235-100	T.19B-16	74-34	70	
CN.235-100	T.19B-17	74-35	74	
CN.235-100	T.19B-18	74-36	75	
CN.235-100	T.19B-19	74-19	76	
CN.235-100	T.19B-20	74-38	79	
Remark: The withdrawn T.19B-03 was sold to Jordan on 19 February 2011.				

CASA C.295M (T.21)

This is a medium tactical transport aircraft that was designed by CASA in 1990 as a development of the CASA-Nurtanio CN.235. From 2000, with the incorporation of Construcciones Aeronáuticas into the European EADS group, the aircraft was designated the EADS CASA C.295. It made its first flight on 28 November 1997 and entered service with the EdA in 2001.

The C.295 can perform a wide variety of missions effectively: tactical and logistic transport, paratrooper and cargo drops, medical evacuation, and also has a maritime patrol version called the C.295 Persuader.

Its designation as C.295 follows the scheme of the aircraft designed by CASA, with the initial C of the manufacturer followed by a three-digit number, the first of which indicates the number of engines, two in this case, the next two being chosen here because of the target payload: 9.5 tonnes.

It has been acquired by several countries in addition to Spain, including Egypt, Poland, Canada, Brazil, Mexico, and Portugal. It has participated in numerous international missions, including missions in the former Yugoslavia, Afghanistan, Iraq, Lebanon, and Chad.

In Spain and as part of the FATAM Programme, in 1998, the EdA decided to purchase nine aircraft to be incorporated from 2001 onwards.

Five more of the same type were acquired between 2006 and 2007.

The 14 aircraft are deployed at Getafe AB.

REGISTRATION	UNIT INDICATIVE	CONSTRUCTOR NUMBER	CRASHED
T.21-01	35-39	S-002	

Dassault Falcon 900 T.18.12, 45-41. (Author's Photo)

CASA CN.235-100, T.19B-11, 35-29 (D.4-07). (Author's Photo)

T.21-02	35-40	S-003	
T.21-03	35-41	S-004	
T.21-04	35-42	S-005	
T.21-05	35-43	S-006	
T.21-06	35-44	S-007	
T.21-07	35-45		
T.21-08	35-46		
T.21-09	35-47		
T.21-10	35-48		03.04.2019
T.21-11	35-49	032	
T.21-12	35-50		
T.21-13	35-51	046	
T-21-14	35-52		

Airbus.A 310-304 (T.22)

The Airbus A.310 is a wide-body commercial aircraft designed and manufactured by Airbus Industries, the European consortium formed at the time by the United Kingdom, France, and West Germany. Airbus had a demand for an aircraft somewhat smaller than the A.300, its first twin-engine wide-body aircraft. The A.310 prototype made its first flight in April 1982. It was certified in March 1983 and entered service with Swissair in April of the same year. Until the last delivery in June 1998, two hundred and fifty-five aircraft were produced, including a cargo version and a military variant. Known as a multi-role transport aircraft, designated A.310 MRTT, it was later converted to a tanker.

The EdA needed a more modern aircraft with greater capacity for VIP missions. In 2001, it purchased two Airbus A.310-304 for the 45th Air Forces Group, which were originally scheduled for delivery in mid/late 2002, but due to technical and fitting-out problems, were not delivered until a year later.

The two aircraft came from Air France and were assigned the constructor's numbers 550 and 551.

REGISTRATION	UNIT INDICATIVE	CONSTRUCTOR NUMBER
T.22-01	45-50	550
T.22–02	45-51	551

Cessna Citation V (TR.20)

In October 1968, Cessna announced its intention to manufacture its first jet aircraft, which was initially designated 'Fanjet 500'. The prototype first flew on 15 September 1969, by which time the original designation was changed to the Model 500 Citation. The second prototype joined the test programme in January 1970, with deliveries beginning in September 1971 after certification. By early 1974, more than 130 units had been delivered, with production running at a rate of 11 units per month.

In the case of the EdA, the first of the two Cessna Citation V acquired to complement the CASA C.212s of the 403rd Squadron (EdA's Cartographic and Photographic Centre) arrived at the Cuatro Vientos air base on 23 December 1992.

In March of the following year a second aircraft arrived.

A third aircraft of this type was later purchased.

The three aircraft are intended for carrying out aerial photography and mapping missions, and for remote detection of targets and objectives.

REGISTRATION	UNIT INDICATIVE	CONSTRUCTOR NUMBER
TR.20-01	403-11	560-0161
TR.20-02	403-12	560-0193
TR.20-03	403-21	

Beech Aircraft C.90 King Air (E.22)

Beech developed the King Air, with pressurisation system, through the military UN-8F powered by PTGA turboprops. The commercial prototype first flew in January 1964, and production of the Model 90 with PTGA-6 engines began in the same year. In 1966, the A.90 model appeared with PTGA-20 engines. Somewhat later, the model evolved and due to the changes incorporated became the C.90. This model has a wingspan of 15.32 metres, a length of 10.82 metres and a total weight of 4,350 kilograms.

In 1974, six such aircraft were purchased and served in the 744th Squadron. In 1975, another four were bought and also flew in the same Squadron.

Later, all of them were transferred to the National School of Aeronautics, where they flew with civilian registrations until 2002, when five of them returned to the EdA with renewed registrations: U.22-01 to U.22-05.

REGISTRATION	UNIT INDICATIVE	CONSTRUCTOR NUMBER	CRASHED
E.22-01	42-30	LJ-666	
E.22-02	42-31	LJ-621	
E.22-03	42-32		18.02.2004
E.22-04	42-33		
E.22-05	42-34	LJ-605	

Beech Aircraft F-33A Bonanza (E.24A)

Despite the decades that have passed since Beech flew the prototype of this light aircraft (September 1959), the aircraft remains one of the most popular and attractive. It is also true that in more than 70 years, the manufacturer has introduced many improvements, which has allowed it to remain, within its category, a basic teaching aircraft.

In 1974, the EdA ordered a batch of twelve Bonanza F-33A from Beech, which was shortly followed by a second order of 18, all of them destined for the 791st Squadron of the General Air Academy, based in San Javier, and the 421st Squadron of the 42nd Group in Getafe, the latter dedicated to the retraining of pilots who fly occasionally, a main mission that they combine with liaison and small transport missions.

REGISTRATION	UNIT INDICATIVE	CONSTRUCTOR NUMBER	WITHDRAWN	CRASHED
E.24A-1	42-01	CJ-52		
E.24A-2	42-02	CJ-54		
E.24A-3	42-03	CJ-55		
E.24A-4	42-04	CJ-57	n/i	
E.24A-5	42-05	CJ-58		
E.24A-6	42-06	CJ-59		
E.24A-7	42-07	CJ-50		
E.24A-8	42-08	CJ-61	n/i	
E.24A-9	42-09	CJ-67		
E.24A-10	42-10	CJ-68		

E.24A-11	42-11	CJ-75		
E.24A-12	42-21	CJ-76		
E.24A-13	42-12	CJ-111		
E.24A-14	42-22	CJ-112		
E.24A-15	42-13	CJ-113		
E.24A-16	42-23	CJ-114		
E.24A-17	42-14	CJ-115		22.09.1993
E.24A-18	42-24	CJ-116		
E.24A-19	42-15	CJ-117		
E.24A-20	42-25	CJ-118	n/i	
E.24A-21	42-16	CJ-119		
E.24A-22	42-26	CJ-120		
E.24A-23	42-17	CJ-121		
E.24A-24		CJ-122		1978
E.24A-25	42-18	CJ-123		
E.24A-26	42-19	CJ-124		24.01.1997
E.24A-27	42-20	CJ-125		
E.24A-28		CJ-126		14.11.1984
E.24A-29	42-27	CJ-127		
E.24A-30	42-28	CJ-128	n/i	

CASA C.101EB Aviojet/Mirlo (XE.25 and E.25)

A Spanish representative of the second generation of basic training jets, the initial C.101 project dates back to mid-1974, a time when the Spanish aeronautical industry was going through the crisis of the cancellation of the CASA C.401 medium transport project. Approved by the Government in January 1975, the C.101 programme went from a simple proposal to the drawing board. The objective was to obtain a basic and advanced training jet to enter service throughout the 1980s and 1990s that would provide Spanish military aviation with a suitable replacement, not only for the Hispano Aviación HA.200 Saeta and Lockheed T-33A, but also for the North American T-6 Texan. Among other things, the new aircraft was required to be highly manoeuvrable, energy efficient and easy to operate.

The first prototype flew in June 1977, meeting the short development times required by commercial export requirements. This first aircraft was part of the four prototypes plus a structure that acted as a test bed. Once tested by CASA personnel, these four aircraft were handed over to the EdA's 406th Flight Experiment Squadron for evaluation.

During 1978, the newly created Ministry of Defence contracted a total of 17 units, which later became 88. In the meantime, after the necessary tests carried out by 406th Squadron, the aircraft was declared suitable for basic and advanced training.

The first production aircraft made its maiden flight in October 1979 and after this event, in 1980, units began to be delivered to the 793rd Squadron of the AGA. Other aircraft of this type were assigned to the 74th Group and later some of them were incorporated – armed – into the 21st Wing at Morón.

The four prototypes were designated by the manufacturer as EB01, as well as the 88 production aircraft.

In due course, 14 of them became part of the 'Patrulla Águila', the EdA's official aerobatic patrol. For this mission, they were equipped with the fumigant elements.

Part of the C.101 Aviojet, or Mirlo, as they are also known, have been decommissioned, whilst the rest will also be decommissioned when the Pilatus PC-21s are delivered, which will start to happen in 2022.

The numbering jumps that the reader will notice in the constructor's (CASA) number list are due to the fact that the C.101s sold to Jordan, Chile and Honduras were taken off the assembly line.

The C.101 Aviojets will be withdrawn from service in mid-2022 when new Pilatus PC-21s enter service and take over the 2022-2023 AGA training course.

Cessna Citation V, TR.20-02, 403-12. (Author's Photo)

Beechcraft F-33C Bonanza, E.24A-11, 42-11. (Author's Photo)

REGISTRATION	UNIT INDICATIVE	CONSTRUCTOR NUMBER	WITHDRAWN	CRASHED
PROTOTYPES:				
XE.25-01	406-01	991	31.05.1992 [1]	
XE.25-02	406-02	992	31.12.1992 [2]	
XE.25-03	406-03	993		16.08.1984
XE.25-04	406-24	994	31.12.1992 [3]	
SERIAL AIRCRAFT:				
E.25-01	79-01	1	n/i	
E.25-02		2		09.09.1983
E.25-03	79-03	3		03.10.1995
E.25-04		4		04.02.1985
E.25-05	79-05	5		
E.25-06	79-06	6		
E.25-07	79-07	7		
E.25-08	79-08	8		
E.25-09	79-09	9		
E.25-10	79-10	10		
E.25-11	79-11	11		
E.25-12	79-12	12		
E.25-13	79-13	13		
E.25-14	79-14	14		
E.25-15	79-15	15		
E.25-16	79-16	16		
E.25-17	74-40	17		
E.25-18	74-42	18		
E.25-19	79-19	19		
E.25-20	79-20	20		28.08.2019
E.25-21	79-21	21		
E.25-22	79-22	22		
E.25-23	79-23	23	n/i	
E.25-24	79-24	24		
E.25-25	79-25	25		
E.25-26	79-26	26		
E.25-27	79-27	27		
E.25-28	79-28	28		
E.25-29	74-45	29		
E.25-30	79-30	30		30.10.1997
E.25-31	79-31	31		
E.25-32	74-01	32		12.01.1995
E.25-33	74-02	33		
E.25-34	74-44	34		
E.25-35	54-20	35		
E.25-36	79-36	37		30.09.1999
E.25-37	79-37	38		
E.25-38	79-38	39		
E.25-39	79-39	40		26.10.1992
E.25-40	79-40	41		
E.25-41	74-41	42		
E.25-42		43		15.10.1993
E.25-43	74-43	44		
E.25-44	79-44	45		
E.25-45	79-45	46		
E.25-46	79-46	47		
E.25-47	79-47	48		
E.25-48	79-48	49		
E.25-49	79-49	50		
E.25-50	79-33	51		
E.25-51	74-07	52		
E.25-52	79-34	53		

CASA C.101EB Aviojet, E.25-51, 411-06 (74-07). (Author's Photo)

E.25-53	74-09	54		
E.25-54	74-10	55		
E.25-55	54-21	57		
E.25-56	74-11	58		
E.25-57	74-12	59		
E.25-58	54-06	60		13.03.1995
E.25-59	74-13	61		
E.25-60	74-14	62		12.01.1995
E.25-61	74-15	63		
E.25-62	74-16	67		
E.25-63	74-17	68		
E.25-64	74-18	69		
E.25-65	74-19	71		
E.25-66	74-20	72		
E.25-67	74-21	73		
E.25-68	74-22	74		
E.25-69	74-23	75		
E.25-70	74-24	76		10.05.1991
E.25-71	74-25	78		
E.25-72	74-26	79		
E.25-73	74-27	80		
E.25-74	74-28	81		
E.25-75	74-29	83		26.04.2012
E.25-76	74-30	84		
E.25-77	74-31	85		10.05.1991
E.25-78	79-02	86		
E.25-79	74-32	87		
E.25-80	74-33	88		
E.25-81	74-34	89		
E.25-82		91		13.06.1984
E.25-83	74-35	82		
E.25-84	79-04	93		
E.25-85	74-36	94		22.06.1990
E.25-86	79-32	97		
E.25-87		101		
E.25-88	74-39	102		

[1] This prototype held these civil registrations: EC-ZDF; EC-ZDY; EC-ZDZ; EC-ZZU; EC-ZZZ. It is currently in the Air Museum, in Cuatro Vientos.
[2] This prototype held the civil registration EC-ZDZ.
[3] This prototype held the civil registration EC-ZDI. One or more of them held the civil registration EC-DUJ.

Remark: It must be considered that, in addition to the crashed aircraft and therefore removed from this list, other accidents have occurred about which no information was given. These happened as follows:
- September 2005: carried the decoration of 'Patrulla Águila'.
- July 2008: one of them was withdrawn (accident?).
- 27 February 2020: one of them falls into the sea.
- February 2016: E.25-45, 79-45, crashed when landing at San Javier AB. It is unknown whether it was repaired or removed.

The Aviojet that make up 'Patrulla Águila' are those registered: E.25.01, E.25.26, E.25.13, E.25.14, E.25.25, E.25.06, E.25.07, E.25.08; E.25.21, E.25.23, E.25.27, E.25. 28, E.25.22, E.25.40 and E.25.52.

ENAER T-35C Pillán/Tamiz (E.26)

Designed by the Chilean firm Empresa Nacional de Aeronáutica (ENAER) in collaboration with the American firm Piper Aircraft, the Pillán is basically a heavily modified version of the Piper Cherokee for military use. Its incorporation into the EdA was motivated by a commercial operation initiated by CASA as a result of the sale to Chile of the C.101 Aviojet trainer. According to an agreement between the governments of both countries, CASA would assemble a series of 40 kits supplied by ENAER at its Getafe factory which, in turn, would be sold to the EdA as selective training aircraft to carry out the mission previously performed by the old Büker and Mentor, which were withdrawn from service in 1985.

The incorporation of the Pillán was carried out within the framework of the FALES Programme (Futuro Avión Ligero de Enseñanza Selectiva, Future Selective Training Light Aircraft).

The two prototypes were delivered to the EdA in December 1984 and were incorporated into the 406th Flight Experiment Squadron.

The remainder were assembled by CASA in accordance with the schedule.

The first deliveries of the series aircraft took place in July 1987. The first flight with students took place on 12 April 1988. In 1989, they were withdrawn from service after some accidents/incidents, to return to service after some modifications.

With the full incorporation of the Pilatus PC-21s, the teaching work done by the T-35Cs will be completed.

REGISTRATION	UNIT INDICATIVE	CRASHED
E.26-01	79-48, 79-50	
E.26-02	79-49, 54-20, 79-68	
E.26-03	79-50, 79-77	
E.26-04	79-51	
E.26-05	79-52	18.09.2019
E.26-06	79-53	
E.26-07	79-54	
E.26-08	79-55	
E.26-09	79-56	
E.26-10	79-57	
E.26-11	79-58	
E.26-12	79-59	
E.26-13	79-60	
E.26-14	79-61	
E.26-15	79-62	
E.26-16	79-63	
E.26-17	79-64	
E.26-18	79-65	
E.26-19	79-66	
E.26-20	79-67	
E.26-21	79-68	30.04.1991
E.26-22	79-69	
E.26-23	79-70	
E.26-24	79-71	
E.26-25	79-72	
E.26-26	79-73	
E.25-27	79-74	
E.26-28	79-75	
E.26-29	79-76	
E.26-30	79-77	15.01.1992
E.26-31	79-78	
E.26-32	79-79	
E.26-33	79-80	
E.26-34	79-81	
E.26-35	79-82	
E.26-36	79-83	
E.26-37	79-84	
E.26-38	79-85	
E.26-39	79-86	
E.26-40	79-87	14.09.1989

Pilatus Aircraft PC-21

Pilatus Aircraft Ltd was established in 1939 to produce light transport aircraft. Later, in 1945, it began producing two-seat military trainer aircraft for the Swiss Air Force and in the following decades went on to produce a number of models for the international market. In December 2000, the owners of Pilatus sold the company to a consortium of Swiss investors who strengthened the firm, expanded its market, created new models and maintained its prestige by winning a number of contracts covering almost the entire world.

The most recent generation of trainer aircraft is the Pilatus PC-21, which was introduced in 2002. This model, which is powered by a 1,200 hp American Pratt & Whitney Canada PT6A-68B turboprop engine, has achieved sales of 214 examples.

In order to replace the CASA C.101 Aviojet, the EdA undertook the selection of a suitable aircraft for training at the AGA in San Javier.

The selection was made between the Beechcraft T-6 Texan II and the Pilatus PC-21. The aircraft finally chosen was the Pilatus, for which 24 examples, two cockpit simulators and two flight simulators were contracted.

The agreed delivery dates were to start in mid-2020 for completion in September 2021, but due to the Covid-19 pandemic, production was postponed.

The Pilatus PC-21s, which will enter service at the start of the AGA's 2022-2023 academic year, will take over all teaching stages and replace CASA C-101 Aviojets and ENAER T-35C Pillán/Tamiz.

REGISTRATION	UNIT INDICATIVE	CONSTRUCTOR NUMBER
E.27-01	792-01	314
E.27-02	792-02	315
E.27-03	792-03	316
E.27-04	792-04	317
E.27-05	792-05	318
E.27-06	792-06	319
E.27-07	792-07	320
E.27-08	792-08	321
E.27-09	792-09	322
E.27-10	792-10	323
E.27-11	792-11	324
E.27-12	792-12	325
E.27-13	792-13	326
E.27-14	792-14	327
E.27-15	792-15	328
E.27-16	792-16	329
E.27-17	792-17	330
E.27-18	792-18	331
E.27-19	792-19	332
E.27-20	792-20	333
E.27-21	792-21	334
E.27-22	792-22	335
E.27-23	792-23	336
E.27-24	792-24	337

Canadair/Bombardier CL-215, CL-215T and CL-415 (UD.13 and UD.14)

This twin-engine, amphibious Canadian aircraft model was specially designed for forest firefighting, a mission in which it has proven to be highly effective. It can hold more than 5,000 litres of extinguishing mixture or water and, once launched, refuelling can be done in just 12 seconds, whilst the aircraft glides over the surface of a lake, a reservoir, or a calm sea at a speed of only 120km/h. Between the manoeuvres of flying over the surface of the water, loading the liquid, and lifting off, the aircraft usually covers a distance of approximately 1,770 metres.

The first of the aircraft ordered by the province of Quebec (Canada) made its maiden flight on 23 October 1967, and its first take-off from the water on 2 May 1968. Among the first customers for this aircraft were Canada, France, and Spain. It is worth noting that one of the French Civil Defence Service's aircraft carried out 82 water drops with a total of 447,310 litres in a single day in the summer of 1970, a good example of the great capacity of this aircraft in its specialised role.

The CL-215 was designed with a cantilever wing, whilst its fuselage was all-metal and of the boat hull type. Two hanging floats located near the wingtips ensure perfect stability. A retractable tricycle landing gear, housed in both sides of the fuselage, allowed it to land on unprepared terrain.

In 1969, the Spanish Government leased an example to Canadair to evaluate its performance as a firefighting aircraft. In 1971, the Ministry of Agriculture acquired two aircraft which, with civilian registrations and flown by EdA pilots, were administratively assigned to the 803rd SAR Squadron. In 1973, a special unit was created for them: the 404th Squadron, based at Torrejón. In 1974, eight production aircraft were purchased, followed by seven more which were delivered in 1979. At that time, it changed its name to become the 43rd Group (431st and 432nd Squadrons).

Occasionally these aircraft perform SAR missions as well as transport missions.

In 1984, two new aircraft were purchased, whilst on 21 July 1989 it was decided to purchase 10 new aircraft and to re-engine some of the old ones. These were the UD.13-15 to UD.13-20 (except UD.13-18, which no longer existed). The work was carried out in Canada. Of the new aircraft, four of them, which had arrived with the old configuration, also went to the factory, whilst another two were re-motorised at the Albacete workshops.

The last four aircraft, UD.13-27 to UD.13-30, were incorporated with the re-motorisation made in Canada.

Between 2006 and 2008, three new aircraft were added, this time of the CL-415 type, and in 2014 another aircraft was purchased. To differentiate them, these four aircraft adopted the registration numbers UD.14-01 to UD.14-04.

The decommissioned CL-215s, which still had significant flying potential, were sold to various Spanish and Italian companies.

Aerospatiale AS.332B/B1 Super Puma and AS.332M1 Cougar (HD.21 and HT.21)

In the mid-1970s, the French firm Sud Aviation (SA) began to design and develop a twin-engine helicopter, intended not only to meet the specifications of the French Air Force for tactical and logistical transport in all weather conditions, but also for use by other air forces.

It was designed in response to the demands of ALAT (the French Army's light aviation), which needed a medium-sized helicopter capable of operating in all weather conditions, day and night, to replace the Vertol H-21 and Sikorsky H-34 helicopters used extensively in the Algerian war. As a consequence of this conflict, France had developed helicopter transport as an additional tactic. The transport helicopters of the time did barely what was required of them, as they used piston engines.

The French industry had already produced American helicopters under licence and was beginning to produce its own designs. The French Army wanted a modern twin-engine helicopter that was easily accessible for repairs and could be transported by a C-160 Transall, a very robust aircraft with a large payload capacity.

The result was the SA.330 Puma, a helicopter designed to meet the requirements of the French Army and to carry up to 20 soldiers. This meant developing an entirely new helicopter. The first of the two prototypes initially flew in April 1965, and after basic testing, six more production models were built, the last in 1968. The first deliveries of the resulting helicopter, the SA 330B, were delivered to the French Army in 1969.

After France, it was the United Kingdom that adopted it for its armed forces, and later a number of other countries also acquired large series.

Between 1974 and 1978, Spain purchased a total of nine SA.330C/L, HL/ and J series helicopters, all of which were decommissioned and replaced by more advanced models through the years.

Over the years, SA was taken over by Aerospatiale (AS), which continued the development of this type of helicopter, and later became part of the Eurocopter Group. It was during this period that, based on the Puma model, a new helicopter was developed to replace the obsolete SA.330 – the AS.332 Super Puma, a medium-sized, twin-engine utility helicopter with a four-bladed main rotor. It was originally manufactured by Aerospatiale and made its maiden flight on 13 September 1978. Many versions have been produced, including transport, SAR, and anti-submarine warfare missions, and even, forest firefighting.

Between 1982 and March 1984, the EdA incorporated twelve AS 332B Super Pumas, assigned to SAR/CSAR and 402nd Squadron. Those assigned to SAR were incorporated into 802nd and 803rd Squadrons. Between 1990 and 1992, two more of the AS.332B1 version were purchased for 803rd Squadron. Four AS.332M1 were also purchased for 404th Squadron at this time.

The HD.21-3 was assigned with the Unit Indicative 802-01 of the HD.21-1, which was lost in an accident

Airbus Helicopters H-215 (AS.332C1) Super Puma (HD.21)

Airbus Helicopters continues the saga created by its predecessors. Derived from the successful Super Puma family, Airbus Helicopters created the H.215, a standard version designed for aerial and utility work. It features an experienced and robust design with a modern cockpit and autopilot. This helicopter offers excellent performance and payload lift and a proven ability to operate in the world's most hostile environments. The digital autopilot acts on all four axes of the H.215, providing precision and stability for total mission safety.

The EdA purchased, between 2016 and 2019, nine H.215 (AS. 332C1), all of which are intended to carry out Combat SAR (CSAR) missions and are assigned to 802nd Squadron.

TYPE	REGIS-TRATION	UNIT INDICATIVE	CONSTRUCTOR NUMBER
AS.332C-1 (H-215)	HD.21-15	SAR	2057
AS.332C-1 (H-215)	HD.21-16	802-16 SAR	2295

The first jet-powered aircraft to enter production and service in Spain was the Hispano HA.200 Saeta. Designed by German aircraft designer Willy Messerschmitt, Saeta first flew in 1955, but due to problems with acquisition of suitable powerplants, entered series production only in 1962. Deliveries to EdA commenced the same year, with the type receiving the service designation C.10 and E.14 and being used as a jet trainer for two decades. In 1970, HA.200s entered service with Canary Islands-based Escuadrón 462, from where they made frequent forward deployments to the Spanish Sahara. Jets of this unit received this distinct camouflage pattern and saw some combat action against the Polisario Front, in 1974, usually armed with unguided rockets calibre 68mm. (Artwork by Tom Cooper)

The production of the HA.200 was followed by the single-seat version designated the HA.220 Super Saeta. While still powered by the same pair of Turbomeca Marborè turbojets like the HA.200, and retaining the EdA's designation C.10, this variant was custom-tailored for ground attack. The HA.220 made its first flight in 1970 and entered service with EdA a year later. For ground attack, this version received two hardpoints under each wing, used to carry bombs and rocket pods. Overall, CASA manufactured 2 HA.200R prototypes, 30 HA.200As, ten HA.200Bs, 55 HA.200Es, and 25 HA.220 Super Saetas: 90 additional HA.200B al-Kahiras were manufactured in Egypt under licence. (Artwork by Tom Cooper)

On 26 September 1953, Madrid and Washington signed the Spanish-American Defence Treaty, under which the US government agreed to provide modern combat and training aircraft and to train Spanish air and ground crews in exchange for Spain providing bases for bombers of the USAF Strategic Air Command. In early 1954, this resulted in the delivery of 30 Lockheed T-33A Shooting Star jet trainers, most of which were operated by the Escuela de Reacción. In Spanish service, the type received the designation E.15, and corresponding serials and codes – although retaining its 'bare metal overall livery throughout the career. (Artwork by Tom Cooper)

Thanks to the Spanish-American Defence Treaty, starting in September 1955, the EdA received its first 25 North American F-86F Sabre fighter-bombers. Over the following two years, more than 200 additional Sabres followed. They equipped three day-fighter wings, each of which comprised three squadrons of 25 aircraft and thus formed the backbone of what became the Air Defence Command (Aviación de Defensa Aérea) of the EdA, which designated it the C5. In Spanish service, Sabres retained their bare metal overall livery. As obvious from the crest (shown inset), this example served with the Ala 1 of the late 1950s. (Artwork by Tom Cooper)

In 1962-1963, the Air Defence Command of the EdA was reinforced through the acquisition of 19 Lockheed (though partially manufactured by Canadair) F-104G Starfighter single-seat interceptors and 2 TF-104G two-seat conversion trainers with combat capability. They received EdA's designations C.8 and CE.8, respectively. Actually leased to Madrid, Spanish Starfighters had the distinction of never suffering a single accident throughout their career with Ala 6 (Escuadróns 61 and 161), Ala 12 (Escuadróns 104 and 122), and the independent Escuadrón 104 (insignia shown inset). However, they were withdrawn from service already in 1972 and returned to the USA, which then distributed them to Greece and Turkey. (Artwork by Tom Cooper)

In early 1960s, Madrid sought to both replace EdA's F-86F-fleet through 100 advanced interceptors and upgrade its aviation industry. Negotiations with Washington led to the decision to acquire 100 Northrop F-5A/B Freedom Fighter fighter-bombers, 24 percent of which were to be manufactured by CASA. In 1971, the initial order for 36 F-5As and 34 F-5Bs was modified to 18 F-5As, 18 RF-5As, and 34 F-5Bs, which entered service with EdA under the designations C.9/SF-5A, CR.9/SRF-5A, and CE.9/SF-5B. Deliveries of CR.9/SRF-5A reconnaissance jets began in April 1972: at the time they were painted in sliver-grey overall, as shown inset. Later, the entire fleet was supposed to be camouflaged in the USAF South-East Asia pattern but Spanish colours used proved to be significantly different, resulting in the appearance shown on the main illustration. (Artwork by Tom Cooper)

Aiming to replace F-86F Sabres and F-104G Starfighters, in early 1970s EdA acquired 36 McDonnell-Douglas F-4C Phantom II interceptors, which received the local designation C.12. All the aircraft were former mounts of the 81st Tactical Fighter Wing, US Air Force, which used to be based at RAF Bentwaters, in Great Britain. After undergoing overhaul by CASA at Getafe, they entered service with the Ala 12 at Torrejon AB, which around the same time adopted the crest of the Ala 6. Former F-86-unit, 121 Escuadrón, was the first to convert, in March 1971, followed by 122 Escuadrón (former F-104G-unit), in May 1972. Phantoms received serials C.12-01 to C.12-36, and their attrition was replaced by four F-4Cs in October 1978, but EdA began withdrawing the type from service in 1978. Principal armament was as shown here: AIM-7E Sparrow and AIM-9J Sidewinder air-to-air missiles. Optionally, M61A Vulcan gun calibre 20mm (in the SUU-16/A pod) could be installed under the centreline. (Artwork by Tom Cooper)

In October 1978, Spain acquired its first four McDonnell-Douglas RF-4Cs – all former mounts of the 363rd TRW, USAF. Locally designated the CR.12, they received no codes and were operated by a specialised flight of the 12 Ala. Eight additional RF-4Cs (all former mounts of the 123rd TRW, Kentucky Air National Guard), followed: their availability enabled the establishment of the 123 Escuadrón of the 12 Ala. The same unit operated a single RF-4C from the original batch (CR.12-42), and six F-4Cs that served as target tugs and weapons test platforms. While originally wearing the USAF Hill II Schema in two greys (FS36320 and FS36118), Spanish reconnaissance Phantoms were repainted in FS36375 while undergoing overhauls and then updated through the addition of the AN/APQ-172 terrain-following radar, new ECM-systems and other equipment. (Artwork by Tom Cooper)

Aiming to replace HA.200/220s, in 1975, Madrid commissioned CASA to commence the work on a new training jet. Developed in cooperation with Germany's Messerschmitt-Bölkow-Blohm Corporation, this led to the C.101, a single-engine, jet-powered advanced trainer and light attack aircraft. Production began in 1980, and a total of 88 C.101EB-01s were delivered to the EdA, which designated them the E.25. In the 1990s, the survivors received an extensive upgrade, including improved navigation systems. The improved C.101BB-01s and C.101CC-04s were exported to Chile (as T-36 Halcon), Honduras and Jordan. (Artwork by Luca Canossa)

Seeking to replace its large but obsolescent fleet of light and medium transports by a type with short take off capabilities, in the 1960s Madrid commissioned the CASA to develop a new, turboprop-powered model. The result was the C.212 Aviocar, which first flew in 1971. Large-scale export followed the first order by the EdA (which designated the type as the T.12), resulting in the emergence of five different variants including a dedicated civilian version. This example is shown as operated by Ala 37 in the 1990s. (Artwork by Luca Canossa)

Spain began searching for a new fighter-bomber before joining NATO in 1982. In December of the same year, Madrid selected 72 single-seat McDonell-Douglas F/A-18A and 12 two-seat F/A-18Bs and an order was placed in May 1983. First deliveries followed in 1986: sometimes referred to as EF-18A/B, Spanish Hornets received the official designation C.15 and CE.15, and corresponding serials C.15-13 to C.15-72, and CE.15-01 to CE.15-12, respectively. They were maintained by CASA in Getafe (which was overhauling Canadian CF-18s and F/A-18 of the 6th Fleet US Navy). The jets entered service with Escuadróns 151 and 152 at Zaragoza-Valenzuela and Escuadróns 121 and 122 at Torrejon de Ardoz and were declared operational as of 1988-1989. All originally wore the standard camouflage pattern in FS36495, FS36375, and FS35237. This example is shown as armed with a GBU-10 laser-guided bomb, as deployed by Spanish Hornets during the Operation Allied Force, in 1999, by when the fleet was upgraded to the F/A-18C/D standard. (Artwork by Tom Cooper)

In late 1995, Madrid decided to reinforce the EdA's Hornet-fleet through the acquisition of 24 surplus F/A-18A/Bs from US Navy stocks. The first six arrived the same year, all still wearing the 'Aggressor' camouflage pattern shown here: they received serials C.15-73 to C.15-78. The remainder followed at a rate of six per year until 1998. Over the following years, they were overhauled and then subjected to a mid-life update by CASA in Getafe, during which all were overpainted to the standard camouflage pattern. Except AIM-9L/M Sidewinders, AIM-7M Sparrows, and GBU-10s, by the time EdA had acquired 80 Texas Instruments (later Raytheon) AGM-88 High Speed Anti-Radar Missiles (HARMs), one of which is illustrated as installed on the left outboard underwing pylon of this C.15. (Artwork by Tom Cooper)

Facing the constantly growing threat of wild-fires, in the late 1980s Madrid began searching for an efficient and reliable fire-fighting aircraft. Around the same time, Canadair developed the CL-215 amphibian, which promptly attracted multiple export orders. Spain followed in fashion in 1981, placing its first order for eventual 14 CL.215Ts. All were operated by the 43 Grupo, though one was meanwhile retired from service and put on display at the Museu de Aeronàutica y Astrnàutica in Madrid. Four improved CL.415s (one of which is shown here) were acquired starting in 1996. In the Spanish public, 'Canadairs' are meanwhile some of the best-known aircraft of the modern-day EdA, saving thousands of lives and lots of property every year. (Artwork by Goran Sudar)

Developed in the mid-1970s as a medium-sized commercial utility helicopter, the Sikorsky S-76 attracted significant export orders. Spain acquired eight examples of the S-76C variant, powered by two Pratt & Whitney Canada PT6B-36A/B turboshaft engines, in 1991. Locally designated HE.24, the type entered service with the Ala 78, home-based at Armilla, Grenada. Principal purpose of type is instrument flight training, but the unit has meanwhile seen widespread deployment for medical evacuation, and by October 2014 Ala 78 accumulated 50,000 flight hours on it. A ninth helicopter of this type, a single S-76A, was donated to the EdA by Sikorsky, but never served with the EdA: nowadays it is on display in southern Grenada. (Artwork by Tom Cooper)

In 1983, Spain joined Germany, France, Italy, and Great Britain in the Future European Fighter Aircraft program. Following protracted research and development, and although France withdrew in 1985, this resulted in the Eurofighter EF-2000 Typhoon. Starting in 2003, the EdA acquired a total of 19 Tranche 1, 34 Trance 2, 20 Tranche 3A, and 20 Tranche 4 – for a total of 93 aircraft. Large parts of their structure and avionics were manufactured by Airbus Defence & Space (former CASA) at Getafe. Locally designated the C.16, Spanish EF-2000s assumed quick reaction alert responsibilities in July 2008 when the EdA acquired AIM-120 AMRAAM missiles for them. Meanwhile, they are foremost armed with Meteor and IRIS-T air-to-air missiles, as illustrated here on an example operated by the Ala 11. (Artworks by Tom Cooper)

Unit Insignia, Ejército del Aire

Ala 11

Ala 11, 111 Escuadrón

Ala 12

Ala 12, 121 Escuadrón

Ala 12, 122 Escuadrón

Ala 14

Ala 14, 141 Escuadrón

Ala 14, 142 Escuadrón

Ala 15

Ala 15, 151 Escuadrón

Ala 15, 152 Escuadrón

Ala 23

Ala 23, 231 Escuadrón

(Artworks by Tom Cooper)

Unit Insignia, Ejército del Aire

Ala 31

Ala 31, 311 Escuadrón

Ala 31, 312 Escuadrón

Ala 35

Ala 35, 351 Escuadrón

Ala 35, 352 Escuadrón

Ala 37

Ala Mixta 46

Ala Mixta 46, 802 Escuadrón

Ala 48

Ala 48, 402 Escuadrón

Ala 48, 803 Escuadrón

Ala 49

Ala 49, 801 Escuadrón

Ala 78

43 Grupo

(Artworks by Tom Cooper)

Map of the Air Bases and their locations. (Map by Tom Cooper)

Canadair CL-215T, UD.13-27, 43-27. (Author's Photo)

AS.332C-1 (H-215)	HD.21-17	802-17 SAR	
AS.332C-1 (H-215)	HD.21-18	802-18 SAR	10153
AS.332C-1 (H-215)	HD.21-19	802-19 SAR	
H-215	HD.21-21	802-21 SAR	10216
H-215	HD.21-22	802-22 SAR	
H-215	HD.21-23	802-23 SAR	
H-215	HD.21-24	802-24 SAR	

EUROCOPTER AS 532UL COUGAR (HT.27)

The AS.532UL Cougar is a twin-engine medium multi-role helicopter produced by the Eurocopter Group as a development and upgrade of the former Aerospatiale's military AS.332 Super Puma.

The 402nd Squadron EdA has been operating two examples of this elongated version since 2004.

TYPE	REGISTRATION	UNIT INDICATIVE
AS.532UL	HT.27-01	402-24
AS.532UL	HT.27-02	402-25

SIKORSKY S-76 SPIRIT (HE.24)

This is a medium-sized multi-role helicopter designed by the US firm Sikorsky Aircraft Corporation. The S-76 has two engines. The two rotors are four-bladed. The tricycle landing gear is retractable. It flew for the first time in February 1977 and since that time, well over 1,500 helicopters have been built, operating both in civilian companies and in the military environment, the latter on SAR, coastguard, and similar missions.

Its development began in 1975 as a helicopter capable of carrying 14 people.

In December 1990, the EdA purchased eight helicopters of this type to meet the needs of the H/X programme for helicopter pilot training. The helicopters were delivered between November 1991

TYPE	REGIS-TRATION	UNIT INDICATIVE	CONSTRUCTOR NUMBER	WITH-DRAWN	CRASHED
CL-215	UD.13-1	43-01	1010	n/i	
CL-215	UD.13-2		1013		02.03.1981
CL-215	UD.13-3	43-03	1031	1995	
CL-215	UD.13-4	43-04	1032		09.09.1988
CL-215	UD.13-5	43-05	1033	1995	
CL-215	UD.13-6	43-06	1034	11.06.1990	
CL-215	UD.13-7		1035	19.09.1976	
CL-215	UD.13-8		1036		11.04.1977
CL-215	UD.13-9		1037		07.03.1977
CL-215	UD.13-10	43-10	1038	11.06.1990	
CL-215	UD.13-11	43-11	1051	1995	
CL-215	UD.13-12	43-12	1052	1995	
CL-215	UD.13-13		1053		12.10.1980
CL-215	UD.13-14	43-14	1054	1995	
CL-215T	UD.13-15	43-15	1056		
CL-215T	UD.13-16	43-16	1057		
CL-215T	UD.13-17	43-17	1061		
CL-215T	UD.13-18		1079		13.02.1987
CL-215T	UD.13-19	43-19	1080		
CL-215T	UD.13-20	43-20	1113		
CL-215T	UD.13-21	43-21	1116		
CL-215T	UD.13-22	43-22	1109		
CL-215T	UD.13-23	43-23	1117		25.03.2003
CL-215T	UD.13-24	43-24	1118		
CL-215T	UD.13-25	43-25	1119		

Continued on page 38

CL-215T	UD.13-26	43-26	1120		
CL-215T	UD.13-27	43-27	1121		
CL-215T	UD.13-28	43-28	1122		
CL-215T	UD.13-29	43-29	1124		30.03.2003
CL-215T	UD.13-30	43-30	1125		
CL-415	UD.14-01	43-31	2064		
CL-415	UD.14-02	43-32	2067		
CL-415	UD.14-03	43-33	2069		
CL-415	UD.14-04	43-34	2090		

TYPE	REGISTRATION	UNIT INDICATIVE	CONSTRUCTOR NUMBER	CRASHED
AS.332-B	HD.21-1	802-01 SAR	2039	05.11.1992
AS.332-B	HD.21-2	803-02 SAR	2054	
AS.332-B	HD.21-3	802-01 SAR [1]	2031	
AS.332-B	HD.21-4	802-02 SAR	2035	
AS.332-B	HD.21-5	SAR	2051	20.09.1985
AS.332-B	HD.21-6	803-08 SAR	2055	25.09.2007
AS.332-B	HD.21-7	802-03 SAR	2062	22.10.2015
AS.332-B	HD.21-8	803-07 SAR	2065	
AS.332-B	HD.21-9	803-09 SAR	2066	
AS.332-B	HD.21-10	802-04 SAR	2076	19.03.2014
AS.332-B	HT.21-01	803-13	2062	
AS.332-B	HT.21-02 HD.21-14	803-14 CSAR	2081	
AS.332B-1	HD.21-11	803-11 CSAR	2345	
AS.332B-1	HD.21-12	803-12 CSAR	2354	03.08.2012
AS 532M-1	HT.21A-1	402-20 SAR	2326	
AS.332M-1	HT.21A-2	402-21	2329	
AS.532M-1	HT.21A-3	402-22	2362	
AS.532M1	HT.21A-4	402-23	2370	

and May 1993. Its main mission is as a school, although it also performs SAR missions and tactical and logistical transport.

REGISTRATION	UNIT INDICATIVE	CONSTRUCTOR NUMBER
HE.24-1	78-01	760389
HE.24-2	78-02	760390
HE.24-3	78-03	760394
HE.24-4	78-04	760396
HE.24-5	78-05	760401
HE.24-6	78-06	760402
HE.24-7	78-07	760411
HE.24-8	78-08	760412

EUROCOPTER EC.120B COLIBRI (HE.25)

The Colibri is a light, single-engine, five-seat helicopter designed and developed jointly by Eurocopter, China National Aero-Technology Import & Export Corporation, Harbin Aviation Industries and Singapore Technologies Aerospace.

The EC.120Bs were manufactured at the Marignane (France) production plant and in Australia, whilst the HC.120s are manufactured in China by Harbin.

The Colibri's wide-cabin design gives it excellent visibility, making it suitable for a wide variety of civil missions. In its medical evacuation version, the helicopter can carry a pilot, stretcher and one or two medics. It has an external crane that can lift up to 700kg.

It is mainly made of composite materials. The Turboméca Arrius 2F engine drives the two-bladed main rotor and a protected tail rotor called Fenestron, making it silent, agile and fast.

In its military role, it is used as a training, observation, and light transport helicopter.

It was chosen by the EdA to replace the veteran Hughes 269A-1s of the Helicopter School. A total of 15 helicopters of this type were contracted and began to be delivered from June 2000.

In addition to the EdA, these helicopters provide training services to future helicopter pilots of the Spanish Navy's Arma Aérea, the

Aerospatiale AS.332B Super Puma, HD.21-12, 803-02, from the SAR. (Author's Photo)

Armed Police, the Guardia Civil, and the Customs Surveillance Service (SVA).

In the middle of 2004, the 'Patrulla Aspa' was formed with these helicopters, which on request, goes to aerial demonstrations.

REGISTRATION	UNIT INDICATIVE	CONSTRUCTOR NUMBER
HE.25-01	78-20	1105
HE.25-02	78-21	1140
HE.25-03	78-22	1150
HE.25-04	78-23	1166
HE.25-05	78-24	1167
HE.25-06	78-25	1175
HE.25-07	78-26	1176
HE.25-08	78-27	1197
HE.25-09	78-28	1198
HE.25-10	78-29	1210
HE.25-11	78-30	1212
HE.25-12	78-31	1222
HE.25-13	78-32	1223
HE.25-14	78-33	1231
HE.25-15	78-34	1232

NHIndustries/Airbus Helicopters NH.90STD-2 Lobo (HD.29)

The NH.90 is a medium-sized, twin-engine, multi-role military helicopter built by NHIndustries, a company formed by Eurocopter, Leonardo and Stork Fokker Aerospace. This helicopter, which can be manned by a single pilot, was designed to operate in adverse weather conditions, both day and night.

In early September 1992, NHIndustries signed a contract with the NATO Helicopter Management Agency (NAHEMA) for the design and development of the NH 90, representing the four participating countries: France, Italy, Germany, and the Netherlands, later joined by Portugal.

Design began in 1993 and the first prototype, PT1, made its first flight in December 1995. The second, PT2, flew for the first time in March 1997, whilst the third, PT3, flew for the first time in November 1998. The first contract between NAHEMA and NHIndustries to produce 298 helicopters was signed in June 2000. They were all destined for Sweden, Finland, Norway, Greece, Germany, Italy, Portugal, and France.

In November 2018, Spain committed to the purchase of 23 helicopters for its armed forces: six for the EdA and the rest for the Navy (Arma Aérea de la Armada) and the Army.

The EdA helicopters, registered HD.29-16 to HD.29-21, are ready to perform SAR missions, SAO (Special Air Operations) missions and PR missions.

The EdA intends, when the budget is available, to purchase a further six NH90 TTH in its transport version.

TYPE	REGISTRATION	UNIT INDICATIVE	CONSTRUCTOR NUMBER
NH.90STD-2	HD.29-16	803-16 SAR	1425
NH.90STD-2	HD.29-17	803-17 SAR	
NH.90STD-2	HD.29-18	803-18 SAR	
NH.90STD-2	HD.29-19	803-19 SAR	
NH.90STD-2	HD.29-20	803-20 SAR	
NH.90STD-2	HD.29-21	803-21 SAR	

General Atomics Aeronautical Systems Predator B/ MQ-9 Reaper

At the end of 2015, the EdA purchased four Reapers, Block 5, and two deployable stations for ground control, the necessary sensors to operate the UAVs, and communication and support equipment.

At the end of 2019, the first components of the system arrived at Morón AB, which was assembled and moved to Talavera AB, its main base, to be integrated into the 233rd Squadron, although its deployment base will be the Lanzarote Aerodrome (Lanzarote, Canary Islands).

These aircraft are equipped with MTS-B electro-optical sensors (EO/IR) and Lynx multi-mode radars, Block 20A. The ground stations (GSC), Block 10, have satellite data link (SATCOM, Satélite de Comunicaciones) and line of sight (LOS) capabilities.

Its load capacity is 386 kilos (the internal one) and 1,700 kilos (the external one).

The maximum flight altitude is 15,240 metres, and it can operate for 27 hours.

TYPE	REGISTRATION	ENTRY INTO SERVICE
Predator B	NR.05-01	January 2020
Predator B	NR.05-02	January 2020
Predator B	NR.05-03	2020
Predator B	NR.05-04	2020

Thales Group Fulmar

This is a small model (3.1 metres long) and weighs 20 kilos. Its flying altitude is between 3,000 and 4,000 metres and its speed is 150km/h, with a range of between 8 and 12 hours and 800 kilometres.

It can carry a load of four kilos, which allows the installation of several sensors, such as daylight, infra red, and multispectral cameras.

The EdA incorporated two aircraft and a ground control station in November 2019.

Their destination is the Matacán AB which houses the Grupo Escuelas, where the drone school is integrated.

TYPE	REGISTRATION
Fulmar	–
Fulmar	–

AeroVironment Wasp AE RQ-12A

The EZAPAC (Escuadrón de Zapadores Paracaidistas, Paratrooper Sappers Squadron), deployed at Alcantarilla AB, acquired four UAV WASP AE in 2014.

These devices entered service in 2015 under the name NR-03 to be used as tactical means of reconnaissance.

The system allows a range of 50 minutes of flight time and a maximum observation range of up to five kilometres. It reaches an altitude of 150 metres.

TYPE	REGISTRATION
RQ-12A	NR.03-01

Continued on page 40

RQ-12A	NR.03-02
RQ-12A	NR.03-03
RQ-12A	NR.03-04

AEROVIRONMENT RQ-11 RAVEN

The Compañía Española de Técnica Aeronáutica (CETA), which provides technical and logistical support in Spain for the AeroVironment systems, in 2016 modernised the Raven used by the EdA's Air Deployment Support Squadron (Escuadrón de Apoyo al Despliegue Aéreo, EADA), equipping them with new guided sensors, which weigh only 1.9 kilos. The device is 0.9 metres long and has a wingspan of 1.4 metres. It has a range of 10 kilometres and an autonomy of 90 minutes with a speed of 57 kilometres per hour.

TYPE	REGISTRATION
RQ-11	–
RQ-11	–
RQ-11	–
RQ-11	–
RQ-11	–
RQ-11	–
RQ-11	–
RQ-11	–
RQ-11	–
RQ-11	–

EJÉRCITO DEL AIRE AIRCRAFT (LIST UPDATED IN OCTOBER 2022)

AIRCRAFT OR HELICOPTER	PURCHASED	CRASHED	WITHDRAWN	IN SERVICE
C/CE.15 F-18A/B Hornet	96	11	0	85 [1]
C/CE.16 Eurofighter	73	5	1	67 [2]
AE.9 F-5A, RF-5A, F-5B/M	70	29	22	19
E.25 C.101 Aviojet	92	20	5	67
E.26 T-35C Pillán	40	4	0	36
UD.13 CL-215, CL-215T, CL-415	34	8	9	17
T.12B/C/D TR.12A T.12B TM.12D C.212, Serie 100/200	81	7	67	8 [1]
T.19A/B CN.235	20	0	10	10 [3]
T.21 C.295	14	1	0	13
TR.20 Citation V	3	0	0	3
T.18 Falcon 900	5	0	0	5
E.24A F-33C	30	4	4	22
E.22 C.90 King Air	5	1	0	4
HD.21 AS.332C-1 (H-215) Super Puma	9	0	0	9
HD.21 HT.21/21A AS.332B, AS.332M1	18	6	0	12
HT.27 AS.532UL Cougar	2	0	0	2
HE.24 S-76C Spirit	8	0	0	8
HE.25 EC.120 Colibri	15	0	0	15
T.22 A.310-304	2	0	0	2
D.4 CN.235 VIGMA/SAR	8	0	0	8 [4]
T.23 A.400 Atlas	14	0	0	13 [5]
HD.29 NH.90 TTH Lobo	6	0	0	6 [6]
E.27 PC-21	24	0	0	24 [7]
T.24 A.330-202	3	0	0	1 [8]
DRONES:				
NR.5 General Atomics	4	0	0	4
Thales Group Fulmar	-	-	-	-
NR.03 AeroVironment Wasp	4	0	0	4
AeroVironment Raven	10	0	0	10

[1] The F-18A, C.15-76 (serial number 162426), decommissioned in January 2000, was handed over to EADS CASA for testing.
[2] The Eurofighter XC.16-02 was loaned to EADS CASA by Germany for testing (summer 2003). Finished such tests, it was returned.
[3] The six reformed aircraft that were transferred to the SAR (see note 4) with the registrations D.4-01 to D.4-06 have been considered as technical withdrawals.
[4] The six D.4, from 01 to 06, were previously T.19B transport aircraft.
[5] As of December 2021, 11 aircraft had been delivered and three were still to be delivered.
[6] By December 2021, the first six aircraft had been delivered.
[7] By December 2021, the first nine aircraft had been delivered.
[8] These Airbus will be fitted with refuelling systems in the coming years.

4
COMBAT AIRCRAFT WEAPONS

In order to fulfil their main mission, fighter aircraft have the necessary elements that allow them to fight the enemy in the sky or on the surface of the earth, both from an offensive and defensive point of view.

Modern aircraft can carry a very wide range of weapons, which, arranged in different configurations, make it possible to carry out the mission that the command has arranged.

Generally, fighter and attack aircraft basically mount an internal cannon of 20 or 30mm located at the front of the fuselage. At the same time, under that same fuselage, under the wings or at the tip of them, there are stations where a wide range of launchable weapons (bombs and missiles) of different characteristics and uses can be installed, both for air-to-air and air-to-surface attacks.

In the case of air-to-air missiles, these may be designed to shoot down enemy aircraft, at short or long range, or to be used against targets on the ground (air-to-surface).

The combination to be used in each case can be given by the characteristics of the mission, which is carefully studied during its preparation. Thus, we can see that a combat squadron, or even the aircraft itself, carries both missiles and bombs. It is often the case that, for its own defence, the aircraft loads a minimum of two, short range air-to-air missiles at the outermost stations, leaving the bombs, rockets, or heavier missiles, such as those designed to destroy radars, tanks, or ships, for the indoors stations.

Internal Cannons

M61A-1 Vulcan: The modern internal cannons in service with the EdA are installed on board the McDonnell Douglas F-18 Hornet. This is the 20mm M61A-1 Vulcan rotary weapon, which has an ammunition tank with a capacity of 400 projectiles.

Mauser BK-17: The Eurofighter Typhoon is equipped with a 27mm Mauser BK-27 cannon.

Ammunition: At the end of 1997, the EdA incorporated the 20mm M70-A1 type ammunition from Ranfoss (Norway), manufactured in Spain by EXPAL, S.A., into its war reserve armament. It is a projectile with penetrating, fragmentation, incendiary, and explosive characteristics, which were traditionally distributed among different munitions (HE, HEI, API). This means that a 'universal' type of ammunition is now available, which facilitates logistics. Several orders have subsequently been placed to maintain these stocks.

Launcher Armament

Baja Resistencia, Low Resistance (BR) and Baja Resistencia Paracaídas, Low Resistance Parachute (BRP) Bombs

For many years, the EdA had been using the Mk.82, Mk.83 and Mk.84 series of 250, 500, and 1,000kg free-fall bombs respectively, all of which were of American origin.

A wide range of nationally manufactured bombs are currently used to equip the EdA's fighter aircraft (F-18 Hornet and Eurofighter). These are the multi-purpose devices which, at the request of the EdA, have been designed by INTA (Instituto Nacional de Técnica Aeroespacial, National Institute for Aerospace Technology) and manufactured by EXPAL (Explosivos Alaveses).

These are the BR (Baja Resistencia, Low Resistance) and BRP (Baja Resistencia Paracaídas, Low Resistance Parachute) bombs.

These BR series devices, of which there are models BR.50, BR.125, BR.250, BR.375, BR.500 and BR.1000 – digits that correspond to their calibre in kilos – can be used from different heights and different speeds by the carrier aircraft.

Those equipped with braking parachutes (BRP, braked, and BRPS, super-braked), in their 250 and 500 kilo versions, consisting of a standard BR bomb but equipped with special tails, allow the aircraft to be launched close to the ground and at moderate speeds. The parachute action ensures the aircraft is far away at the time of the explosion.

There is a fragmentation version with a warhead covered with shrapnel.

The EdA has Mk.82, Mk.83 and Mk.84 series weapons of 250, 500, and 1,000 kilos respectively. All these can now be supplemented with Paveway II guidance kits. There are inert versions of these for training purposes.

More modern are the GBU-48 (Enhanced Paveway II) incorporated in 2016 and also called GBU-10 and EGBU-16 (Enhanced GBU-16), manufactured by Raytheon. They are laser-guided bombs and mechanical flight means, which have also received a GPS system. Thus, this guided bomb, manufactured from a 1,000 pound (454 kilo) Mk.83 type bomb, with a range of 15 kilometres, can be accurately employed on targets in all weather conditions, both day and night.

The GBU-10, which is the result of combining a Paveway II guidance system with a conventional Mk 84 bomb, is still in use.

Bombs BPG-2000

This is a precision and high-penetration guided bomb, the result of a programme developed by EXPAL at the request of the EdA. It consists of a high-explosive penetrator head and an electronic delay fuse. Operated by a Paveway III laser guide head.

It was conceived to be used specifically against bunkers to cause devastation inside them.

Practice Bombs

The models BE-6A and BE-11C/A are used for launch exercise.

Missiles

The EdA currently employs a wide range of air-to-air, air-to-ground, and anti-ship missiles, all of which are of American and European origin.

As these are extremely specific devices, the combinations on board the aircraft are quite limited, so they generally carry one pair of air-to-air for their own defence and another pair of offensive missiles intended for the main combat mission that has been scheduled.

AIM-9 Julie and AIM-9L/I Sidewinder

The current range of these infra red and short range guided missiles, in use on the F-18 Hornet and Eurofighter Typhoon, consists of the AIM-9Lima and Lima Improved versions, and the AIM-9 Julie, the improved version of the AIM-9L. The AIM-9N3 version is also used.

AIM-7M and AIM-7P Sparrow

These radar guidance missiles manufactured by Raytheon are in service on units equipped with F-18 Hornets, after many years of experience on F-4C Phantoms. This is a highly effective medium-range air-to-air missile, which the Hornets place in the subalar stations.

AGM-65G Maverick

For the F-18 Hornet, the EdA maintains the AGM-65G model, which is one of the most modern versions of the Maverick family. This is a model that has an infra red imaging optical system with great capacity to acquire terrestrial targets, both day and night.

AGM-84D-1 Harpoon

This is a proven anti-ship missile, developed in the United States by McDonnell Douglas. The EdA uses these weapons on board the F-18 Hornet and Eurofighter Typhoon.

The Armada (Spanish Navy) mounts double or quadruple launchers on all frigates and corvettes in service.

AGM-88B HARM

HARM is the acronym for High-Speed Anti-Radar Missile, which means that it is a tactical air-to-surface device that, in conjunction with the launcher's avionics, detects, identifies, locates, and destroys enemy radar emission sources. The EdA uses the AGM-88B Block III in the F-18 Hornet.

AIM-120B and AIM-120C7 AMRAAM

It can be estimated that the F-18 Hornet and Eurofighter Typhoon have a total of approximately 100 AMRAAM (Advanced Medium-Range Air-to-Air Missile).

AIM-2000 IRIS-T

This is the Infra Red Imaging System Tail conceived and developed by a group of European countries, which is intended to replace the AIM-9 Sidewinder air-to-air missiles. It is a last generation, infra red device with the impressive self-defence capabilities of the F-18 Hornet and Eurofighter Typhoon.

Taurus KEPD-350

Following its decision to join the EdA in 2005 as part of the ALAD Programme (Arma Lanzada a Distancia, Remote Launched Weapon), the Taurus is a small cruise missile intended for the F-18 Hornet and Eurofighter Typhoon. It is a stand-off device, of which there are around 40 units.

M39A-2, 20-millimetre internal cannon from a CASA-Northrop F-5. (Author's Photo)

Pair of DEFA-553 30-millimetre internal cannons. (Author's Photo)

The Future: Meteor

A group of partner companies are working on the project and development of a medium/long range (over 100 kilometres) air-to-air missile called Meteor. Several Spanish companies are involved in this development for the Eurofighter Typhoon.

In 2022, one hundred units of this missile are reported as being on order for delivery in the coming years..

Air Bases Defence Missiles

In recent times the EdA has incorporated several types of missiles which, placed on the ground, have the capacity to defend their air bases.

Integrated Air Defence System

The incorporation of the SHORAD system (Short Range Air Defence System) was approved in September 1995 and the Crotale VT-1/Mistral and Aspide 2000/Mistral combinations were now considered to meet the requirements perfectly.

The final operational and economic evaluation determined the choice of the Aspide missile from the Italian firm Alenia, together

500kg practice bombs under a Hornet. (Author's Photo)

with the Crotale missile from the French firm, Thomson.

In April 1996, the first element of the integrated low and very low-level, air defence system was received at Zaragoza AB and put into service in the Air Deployment Support Squadron (EADA, now SEADA).

This first element consisted of two Mistral (infra red) missiles on each launcher and would, in the future, be completed with Aspide (electromagnetic) missiles and early warning sensors associated with a coordination centre, automatically integrated at all levels of the EdA's Command, Control and Communications (C3) system.

In the foreground, four AIM-9J Sidewinder missiles, in the background, four AIM-7 Sparrow II missiles. (Author's Photo)

Double Atlas Mistral Batteries

These batteries are equipped with Mistral short range infra red missiles, integrated into the Spada 2000 System by means of communications equipment. They are mounted on VAMTAC vehicles (Vehículo de Alta Movilidad Táctico, High Mobility Tactical Vehicle).

Aspide 2000

Each fire section is controlled from a control centre that integrates a tracking and lighting radar and two Aspide 2000 single or double launchers.

They are operated by the SHORAD section of the EADA.

Spada System

Its mission is to provide air defence from the ground to EdA's air bases or other units assigned to it.

In addition to the system's own Aspide 2000 electromagnetic missiles, it includes Atlas Mistral firing positions. They are operated by the SHORAD section of the EADA.

Combat SAR Helicopters Armament

The helicopters adapted for the Combat SAR (CSAR) and Personnel Recovery missions have 12.70-millimetre FN machine gun model M3M holders for these specific missions, replacing the old 7.62-millimetre GIAT F1.

The rescuers operate with Heckler & Koch G-36 assault rifles.

These helicopters have been provided with improved shielding and complex self-protection systems, heat dilution and sand filters for the turbines.

ELECTRO-OPTICAL SYSTEMS FOR COMBAT

FLIR AN/AAS-38 'Nite Hawk'

The F-18 Hornets have been fitted with Ford Aerospace's FLIR/LTDR (AN/AAS-38 'Nite Hawk'). The 18 units purchased are used by the 12th and 15th Wings. At first the FLIR (Forward Looking Infra Red) cameras were received, and the following year, the LTDR (Laser Target Designator and Ranger) capacity subsets arrived.

The aircraft carrying this system is capable of autonomously launching laser-guided bombs. In other words, the same plane that launches the bomb carries the designator with the consequent advantage of not needing the participation of another support aircraft. In the near future, they will be modernised.

FLIR 'LITENING II'

As a result of the experience gained by the EdA during the Kosovo war, it was decided to purchase twenty-two FLIR LTD/R-LST

AIM-9L Sidewinder missile on a Hornet wingtip. (Author's Photo)

'Litening II' units from the Israeli company Rafael, for the F-18A. This is an advanced target designation and navigation system, which can operate both during the day and at night and in adverse weather conditions. These pods came into service during 2001.

FLIR AN/AAQ-22 Safire

To enhance the P-3 Orion maritime patrol aircraft as much as possible, an AN/AAQ-22 Safire thermal imaging sensor has been installed under the radar radome of all of them. These instruments, which are retractable, are housed between the bow radome and the undercarriage housing.

Others

Some F-5Bs retain LTDS AN/AVQ-27 laser designators. These are first generation equipment that were installed in the rear cabs.

Electronic Warfare and Self-protection Equipment

Under this name we group together the equipment and means currently in use in EdA's aircraft, designed to acquire data and provide self-protection to the aircraft when they fly in combat missions.

Although these systems have, in principle, only been installed in the F-18 Hornet and the Eurofighter Typhoon, as budgets permit and the programmes are completed, they will also be included in other types of aircraft and helicopters.

To list them we have grouped these systems from the carrier aircraft:

F-18 Hornet

AMES-P CORE (Capacidad Operacional de Reconocimiento Electrónico, Electronic Reconnaissance Operational Capability) electronic warfare equipment. Manufactured by Indra for the F-18 Hornet, it is a pod for ESM (Electronic Support Measures) and ELINT missions.

Aircraft's own equipment:

- Interferer AN/ALQ-126B
- Interferer AN/ALQ-162 (V) (more modern than the previous one)
- Warning system AN/ALR-67
- Flare and chaff launcher AN/ALE-39

Airbus A 400 Atlas

This large transport aircraft has been equipped with an advanced self-protection system developed by the Spanish firm Indra. This is the InShield infra red countermeasure system of the DIRCM type (Directional Infra red Counter Measures), which will protect these aircraft against ground-to-air missile attacks. The launch of the portable short range missiles guided by a thermal sensor triggers the aircraft's InShield system, which performs a confirmation of the threat by activating the system. The countermeasure is completed by the emission of a laser beam that confuses the missile's guidance system by diverting it from its trajectory. The whole process takes only a few seconds. InShield can act simultaneously against several attacking missiles.

Reccelite

In October 2000, fourteen Reccelite nacelles were acquired from the Israeli firm Rafael, to be operated by F-18A Hornets on tactical and strategic reconnaissance missions. The first of these was delivered in October 2001. They are capable of visual (photographic) and infra red reconnaissance.

Night Vision Glasses

To carry out night missions, air defence units have a certain number of night vision glasses (GVN) type AN/AVS-9 (ANVIS). They can fly at low altitude, even in a tactical formation, and attack targets – even on moonless nights. These glasses are also part of the equipment used in helicopters carrying out Combat SAR missions (CSAR).

For the members of EZAPAC, night vision equipment has recently been purchased.

PHOTOGRAPHIC RECONNAISSANCE EQUIPMENT

RAPTOR

In mid-1999 the EdA announced the purchase of two photo reconnaissance pods for the F-18A starting in 2002. They are of the type called RAPTOR (Reconnaissance Airborne Pod for Tornado), of British origin.

Aerial Photography

In December 2016, the Cartographic and Photographic Centre (CECAF) received the new equipment that makes up the ULTRACAN EAGLE Digital Aero Photogrammetric System, from the GTB Ibérica company, with which to replace the obsolete equipment.

Recon Intelligence Centre

In 1997, the EdA acquired a Reconnaissance Intelligence Centre (RIC) from the US Air National Guard. This facility, transportable

Mistral double launcher for air base defence. (Author's Photo)

in the hold of a former C-130 Hercules, has been used on many occasions in combat exercises. This centre receives the photographs from the aircraft and air reconnaissance systems and once they have been processed, they are sent to the communications network of MACOM headquarters, which then transmits them.

It can now be carried on one of the new A400 Atlas.

Scorpion Helmets for F-18 Pilots

With the addition of the modern ARIS-T missiles, the EdA has equipped the crews of the F-18 Hornets with the modern Thales Scorpion 2 (HGU-55) helmets, which have a firing visor. These helmets are also compatible with the Israeli Rafael Litening pod.

5
AEROSPACE SURVEILLANCE AND CONTROL SYSTEMS

Air Defence – Air Surveillance Squadrons (Escuadrones de Vigilancia Area, EVA)

The current Spanish air surveillance coverage network in 2022 is considered sufficient since, in addition to the real coverage of the radar stations, information on aircraft that may enter Spanish air space is transmitted by neighbouring countries, with which Spain has an agreement on this matter.

However, the concept of 'sufficient' cannot be taken in an absolute sense, as it must be related to the type of threat to which it must correspond.

The constant technological advances in aircraft and missile design are a new threat that needs to be addressed with new technology applicable to air defence.

This requires constant modernisation of air defence radars. Over the years, the EdA has carried out the 'Combat Grande III' and 'Combat Grande IV' programmes to modernise the concepts and technical resources used to update the systems.

The name 'Combat Grande' was equivalent to the Spanish SADA (Sistemas Semi-Automáticos de Defensa Aérea, Semi-Automatic Air Defence Systems).

In 1957, in accordance with the contents of the 'Agreement on Mutual Defence Assistance' (1953) between the United States and Spain, EVAs with an AN/FPS-20 long range radar, an AN/FPS-6 high altitude radar and a TPX-42 secondary radar, were put into service along with the associated systems belonging to the MAP programme (US Military Assistance Programme).

Between 1969 and 1973 and in the 1990s, these types of radar were replaced by the more modern ones: AN/FPS-113 and AN/FPS-90.

Indra, formerly CESELSA, has been replacing this equipment by the ones developed and built in Spain within the framework of the 'Programa 3D LANZA'. Italian equipment from the firm Alenia was installed in some EVAs, specifically the RAT-31 SL-T systems, which are currently being replaced by Indra's LANZA.

The LANZA constitute the Airborne Command and Control System (Sistema de Mando y Control Aéreo, SIMCA).

Remarks:

The EVAs in the Iberian Peninsula depend operationally on the MACOM and organically on the territorial commands.

EVA-6 does not exist. It was installed in Elizondo (Navarra).

The radio code 'Necora' was formerly 'Derby'.

In 2002 there was talk of installing a new site in Pico del Lobo, Cerezo de Arriba (Segovia). Only the road was built.

Remark: The EVA-21 was previously the EVA-8.

In 2022, LANZA equipment are in service in the following EVAs:

EVA-1
EVA-2
EVA-3
EVA-4
EVA-7
EVA-10
EVA-11
EVA-13
EVA-21

The ALENIA RAT-31 SL/T units are in service in the following EVAs:

EVA-5
EVA-9
EVA-12
EVA-22

The four Alenia radars are being replaced by the LANZA model in the LRR version. This will improve their ballistic missile detection capability. This phase will take place during 2022.

In a second phase, planned from 2026 onwards, the remaining nine radars (the LANZA radars currently in service) are planned to be replaced by the LRR model.

Aerospace command and control systems (GRUCEMAC, GRUNOMAC and GRUALERCON) are also being upgraded to increase their surveillance and control capabilities in line with future needs.

Moreover, in the field of airspace surveillance and awareness, priority continues to be given to the final operational training of the Space Surveillance Operations Centre (COVE, Centro de Operaciones de Vigilancia Espacial), scheduled for 2023. This project is proceeding as planned.

Mobile Air Control Group (Grupo Móvil de Control Aéreo, GRUMOCA)

This unit was originally established at Tablada AB, as a development of the Aero-Tactical Control Squadron of the First Transmission Group (Escuadrilla de Control Aerotáctico del Primer Grupo de Transmisiones), which was based in Getafe.

It had been in tactical aviation since the years when the USAF had just transferred the 'Alert and Control Network' to the EdA.

From the 1990s and with the beginning of the EdA's participation in international missions, the GRUMOCA became an essential element aimed at maintaining communications with air detachments established all over the world.

In the 1990s, GRUMOCA began its participation in NATO's 'Command and Control Programme', which required a profound renewal of the technical means at its disposal.

In 1999, Command and Control System Headquarters (Jefatura del Sistema de Mando y Control) was created, attached to the Combat Air Command (MACOM).

In 2006 the unit began a new process of renewing the means of surveillance, command and control, a process which was completed in 2017 with the entry into service of a high technology mobile radar, the AN/TPS-43M, and an ARS/D Control Centre.

6
OTHER ELEMENTS OF THE EJÉRCITO DEL AIRE

Naval Boats

At the beginning of the 1990s, the EdA bought three boats from Rodman/Polyships to serve the SAR from the Military Aerodrome of Pollensa (Island of Mallorca), the Gando AB (Canary Islands) and the San Javier AB (Murcia).

DESTINATION	CODE	NAME
Pollensa Aerodrome	EdA-01	Pollensa II
Gando AB	EdA-02	Avante II
San Javier	EdA-03	–

Built with reinforced polyester and fibreglass, they have a great capacity for search and rescue with adequate autonomy and speed. They have the capacity to transport people and material, as well as divers with their equipment. Its type of construction allows long stays at sea with a minimum of crew.

Features: length 11 metres; beam: 3.90 metres; maximum speed: 22 knots; range: 300 nautical miles.

They are equipped with two 130 HP TAMD-31 Volvo-Penta engines at 5500 rpm.

Special Installations

The EdA has two specialised facilities for the calibration of fighter aircraft radars.

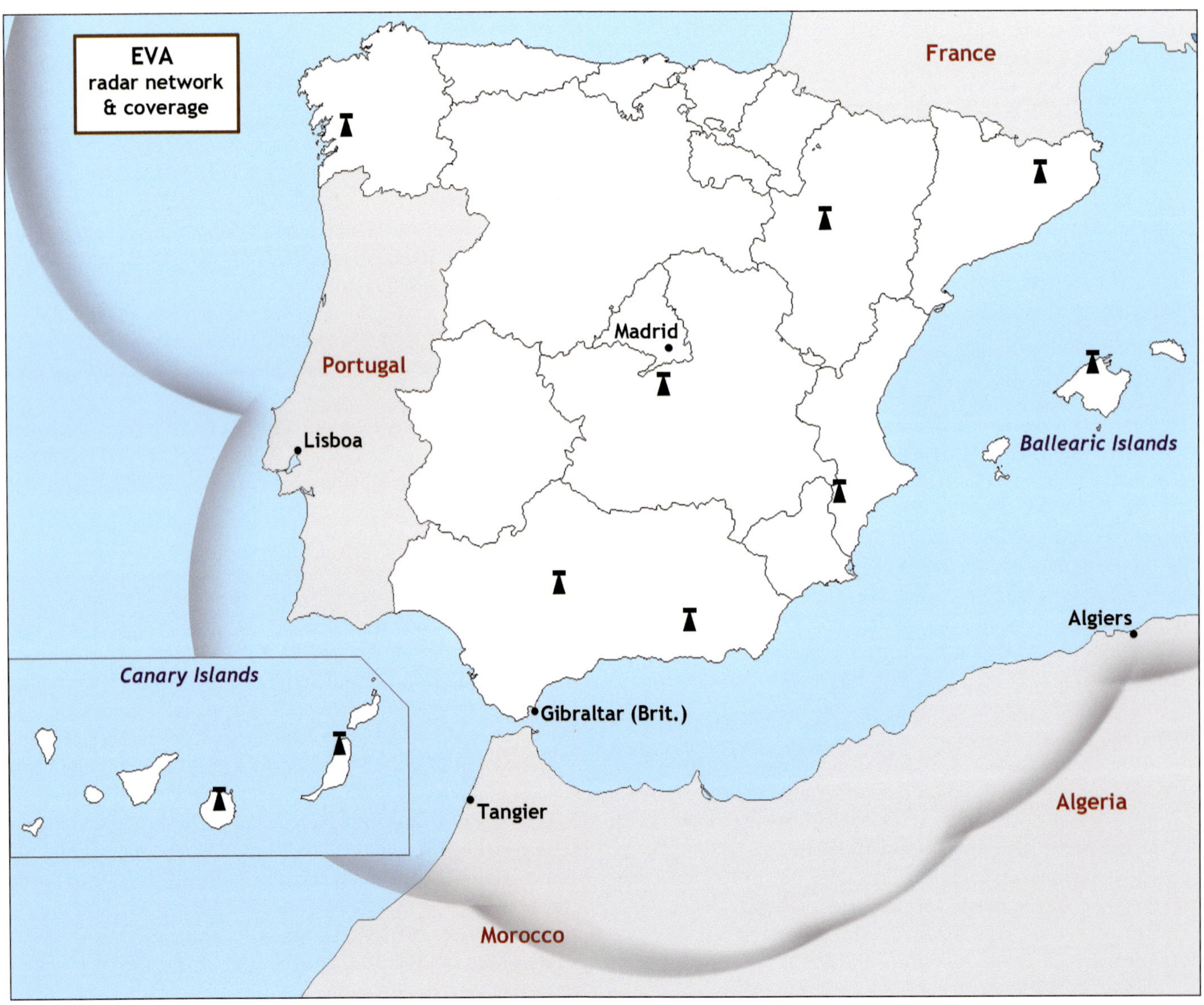

Aerial coverage of EVAs

One is located at Torrejón AB and the other at Albacete AB. They are structurally identical.

Armoured Shelters and Protective Shelters

During the 1980s, the EdA built a series of concrete armoured shelters for the protection of its fighter aircraft stationed at Albacete and Gando Air Bases to protect its main force from possible air attack.

It was said that, initially and in view of the lack of budget, such shelters would not have armoured doors for the full protection of the aircraft in them. Over time, they were only equipped with armoured metal doors that retract inwards, but in any case, offer little protection.

Albacete AB has 12 such shelters, whilst at Gando only seven were built.

At the Morón and Zaragoza Air Bases, considering the high temperatures that they endure for a good part of the year, a large number of tubular metal structures have been built to shelter and protect the Typhoon fighters that would otherwise be in the open.

Flight Simulators and Other Equipment

The flight simulators, or for other uses, available for the EdA are the following:

Two basic tactical and emergency training simulators for the F-18 Hornet, one at Zaragoza AB, 15th Wing, and the other at Torrejón AB, 12th Wing.

Two FMS, Full Mission Simulator, at Morón AB, 11th Wing.

Two ASTA (Aircrew Synthetic Training Aids, CT Cockpit Training) simulators, one at Morón AB and the other at Albacete AB.

Simulators for CASA C.295, CASA C.101 Aviojet, CASA-Northrop F-5M.

MANDO AÉREO DE COMBATE (MACOM)		
GRUCEMAC Torrejón AB Radio Code: PEGASO		
GRUNOMAC Zaragoza AB Radio Code: OSA POLAR		
EVAS		**RADIO CODE**
EVA No. 1	Inoges (Calatayud, Zaragoza)	SIESTA
EVA No. 2	Villatobas (Toledo)	MATADOR
EVA No. 3	Constantina (Sevilla)	BOLERO
EVA No. 4	Monte Paní (Rosas, Girona)	SAMBA
EVA No. 5	Monte Aitana (Alcoy, Alicante)	KANSAS
EVA No. 7	Monte Puig Mayor (Soller, Baleares)	EMBARGO
EVA No. 9	Motril (Granada)	ORION
EVA No. 10	Monte Iroite (Noya, La Coruña)	NECORA
EVA No. 11	Pico del Montero (Alcalá Gazules, Cádiz)	PERSEO
EVA No. 12	Espinosa de los Monteros/Saba (Burgos/Cantabria)	LIRA
EVA No. 13	Sierra Espuña (Totana, Murcia)	DRAGON

A MALE RPA ISR Training simulator (operational simulator) at Salamanca AB, at the UAS (Unmanned Air System) School.

A simulator for Airbus A.400 at Zaragoza AB, in 31st Wing. This is an advanced simulator from Thales.

A simulator (under development) for NH.90 helicopters.

McDonnell Douglas F-18 Hornet

The Spanish company Indra, which at the time developed the flight simulator for the EdA's F-18, was commissioned in 2019 to update the existing simulator in 15th Wing of Zaragoza AB. Indra

Air Surveillance Squadron EVA-2, *Matador* code, from Villatobas (Toledo). (Author's Photo)

MANDO AÉREO DE CANARIAS (MACAN)		
GRUALERCON Gando AB Radio Code: PAPAYO		
EVA No. 21	Pico de las Nieves (Gran Canaria)	NIEVES
EVA No. 22	Peña del Chache (Lanzarote)	SIROCO

has incorporated the training system of the Operational Flight Programme (OFP) 09E version. The objective to be achieved between 2020 and 2021 is for the simulator to accurately replicate the aircraft's functionalities and behaviour to provide the best possible training for pilots.

Air Surveillance Squadron EVA-5 *Kansas* code, Monte Aitana (Alcoy, Alicante). (Author's Collection)

Air Surveillance Squadron,EVA-4, *Samba* code, Monte Paní (Roses, Girona). (Author's Collection)

Air Defence System Control Room, GRUNOMAC (Zaragoza). (Author's Collection)

Eurofighter Typhoon

The Eurofighter training centre is located at Morón AB, in the ASTA simulator department, which in December 2014, completed the first 10,000 hours of training.

Airbus A 400 Atlas

31st Wing of Zaragoza AB has an FFS (Full Flight Simulator), which has been developed by the firm Thales Group, as the main contractor, which has been responsible for the design, development, and production. Indra participated in its construction as a subcontractor. These systems are also in service in Germany, France, and the United Kingdom.

Alcantarilla AB

An advanced positioning air controller simulator (SIMFAC), developed by Indra, has been in operation at this air base since 2013. It is used to train users in Close Air Support (CAS) operations. The

LANZA radar, developed by Indra. (Author's Collection)

Air Surveillance Squadron EVA-21, Nieves code, Pico de las Nieves (Canary Islands). (Author's Collection)

One of the three launches belonging to the SAR. (Author's Photo)

Hardened aircraft shelter at the Gando AB (Canary Islands). (Author's Photo)

Hardened aircraft shelter at the Albacete AB. (Author's Photo)

system recreates the different aircraft guidance missions carried out by the Advanced Air Controller.

Land Vehicles and Other Means of Transport

The EdA has many motor vehicles with which it deals with a multitude of operations related to air bases, aircraft of all types, the transfer of people, elements, and various materials.

Special mention must be made of fire extinguishing trucks, which can act both on the runways, aircraft parking platforms, hangars, as well as in any other installation related to the bases. In the latter case, these are 4x4 and 6x6 trucks.

The Paratrooper Squadron (EZAPAC) has armoured combat vehicles from the firm URO. These are VAMTAC (Vehículos de Alta Movilidad Tácticos, High Mobility Tactical Vehicles) I3 and S3 armoured vehicles, in the basic and extended versions, armed with M-2HB 12.7-millimetre machine guns.

Although they cannot be considered vehicles, as they are not self-propelled, the EdA keeps small devices in service that are used on aircraft parking platforms and in workshops. These include power generators, mobile lighting equipment, missile transporters, bombs, lifting systems and a long list of equipment that is difficult to enumerate.

A pilot from the 15th Wing in the Unit's simulator. (Author's Photo)

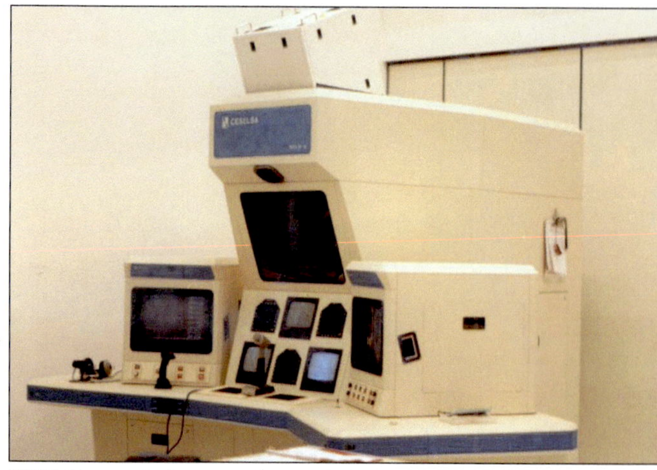

CESELSA, now Indra, manufactured the F-18 Hornet simulators for the 12th and 15th Wings. (Author's Photo)

Cuatro Vientos AB is the home base of the 403rd Squadron. (Author's Photo)

4x4 vehicles

URO VAMTAC armoured version I3 and S3 specific and elongated for EZAPAC, armed with 12.7-millimetres machine guns.
RG-31 Mk 5 Nyala, also for EZAPAC
Nissan Patrol, ML-6, and MC-4 versions
Santana Anibal
Nissan Terrano II, short and long versions
Patrol GR
Nissan Pathfinder
Nissan Navara
Iveco Daily, 4x4 ambulance
Mercedes Unimog truck

6x6 vehicles

IVECO truck M-250.37W

Runway equipment for aircraft operation. (Author's Photo)

Crane for lifting heavy weights and volumes. Secondly, a truck for refuelling. (Author's Photo)

Equipment to start up aircraft. (Author's Photo)

In the maintenance workshop at Cuatro Vientos AB. (Author's Photo)

Auxilliary Power Unit. (Author's Photo)

General Electric, F-404-GE-402 engine, from an F-18 Hornet. (Author's Photo)

Special Vehicles

MB Atego 1225AF, fire extinguisher 4x4 light truck
Protec-Fire MTEC, fire extinguisher 6x6 heavy truck
Protec-Fire MAN LE 14220, fire extinguisher truck

Others

MB-313 CJD Estación Tajuña, for the SAR
Peugeot 407
Peugeot Partner
MB Vito
Iveco Eurocargo
MB Actrol
Renault A296 y 340
Renault Midliner
MB-508
MB-100
MB-511
MB Sprintov
Iveco Eurotracker
Luna RSL-45-9000/A Autocrane
Freightliner M-2106
Pegaso 1216
MB-608D
E. Bike Bultaco Brinco, for the EADA

Shooting and exercise ranges map. (Map by Tom Cooper)

AIR-GROUND:		
SHOOTING RANGE LAS BARDENAS, ACAR LAS BARDENAS		
AIR-AIR:		
LE-26	South of the island of Ibiza	Used for air-to-air shooting against flying darts
SP-3	In the province of Huesca	Used for aircraft-to-aircraft combat
LE-D-34	South of the island of Gran Canaria	Used for air-to-air shooting against flying darts
LE-D-104	In the province of Teruel	It is used for low-level interceptions
LE-D-59	Between Valencia and San Javier	On the coast
LE-D-79	Training area in the south of the Canary Islands (240 x 150 nautical miles)	

7
SHOOTING AND EXERCISE RANGES

Shooting ranges are physical spaces, on the ground, in the air or at sea, where pilots come regularly to carry out aerial exercises and manoeuvres to practise the use of aircraft and weapons and thus preserve their capabilities.

The EdA currently has a large land site located in Las Bardenas Reales (Navarra) on a rental basis whose contract is periodically renewed (for seven years).

Here, Spanish pilots (and those from other NATO nations) conduct live-fire exercises against targets made up of bull's eyes, old equipment (vehicles and aircraft) and buildings constructed for the purpose.

The enclosed areas at sea are also used for live-fire and ship combat simulation.

Las Bardenas Shooting Range. (Author's Photo)

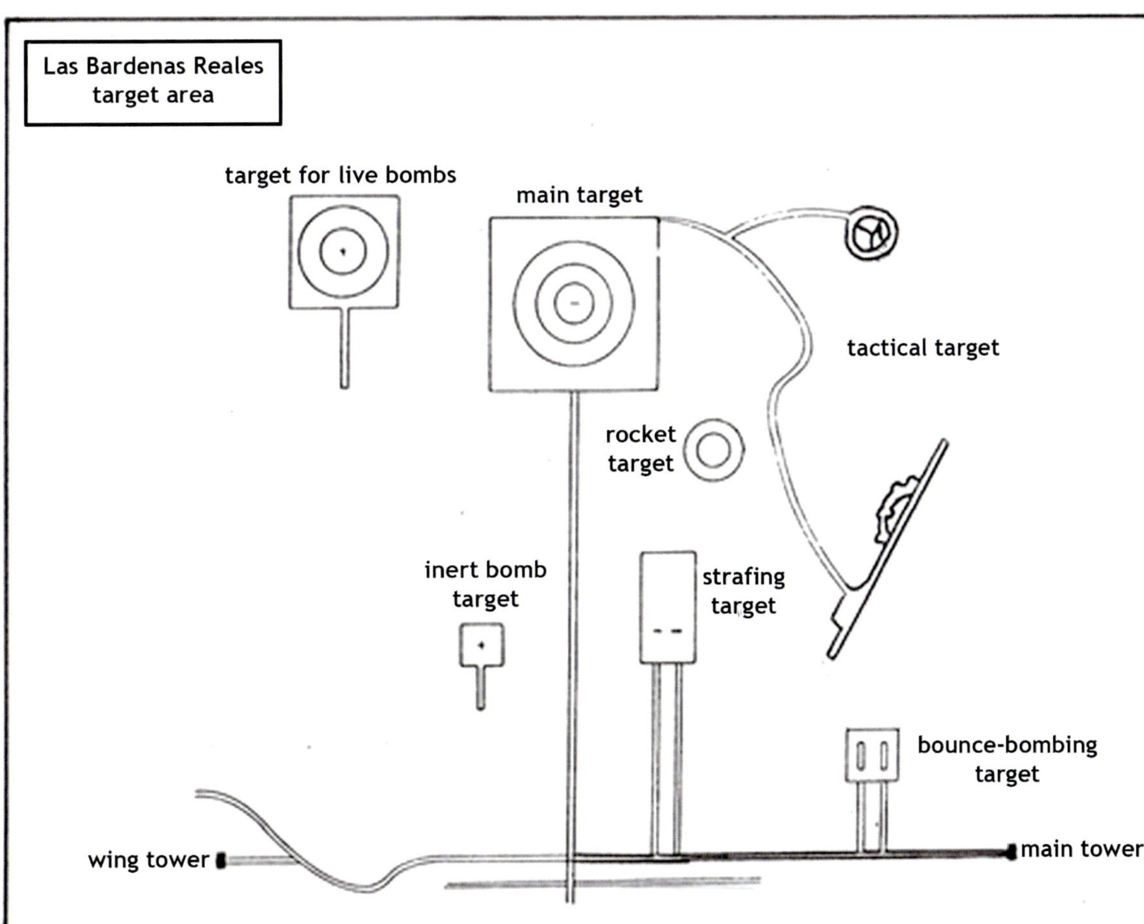

Bardenas Reales target area map.

The areas for combat simulation and aerial manoeuvres are in zones of the Spanish sky, which are activated according to needs. Activity in sea and air areas is warned, both to air and sea navigation, so that they are avoided by ships and aircraft not involved in the exercises.

GEOGRAPHIC LOCATION

Las Bardenas Reales Shooting Range

About 20 kilometres from Tudela (Navarra) and 20 kilometres from Zaragoza, there is an area of 415 square kilometres, arid or sub-desert, known as Las Bardenas Reales.

In 1940, an EdA Captain was commissioned to locate a suitable site in northern Spain for an air-to-ground shooting range. On 13 June 1951, by agreement between the Community of Las Bardenas Reales and the EdA, the range was established. In 1966, the Bardenas Reales went from a single bombing target to the current seven: real bombs, inert bombs, rockets, machine guns, rebound bombs and a tactical target (this was built in 1972), and has provided pilots with a target area against SAM settlements, land vehicles, aircraft, fuel depots and others.

8
OTHER FACILITIES

EZAPAC, Escuadrón de Zapadores Paracaidistas del Ejército del Aire (Paratrooper Sappers Squadron)

Operationally depends on: MACOM (Combat Air Command)
Organisationally depends on: MAGEN (General Air Command)

In 1946, the First Paratrooper Flag of the 1st Airborne Legion (Primera Bandera Paracaidista de la 1 Legión de Tropas de Aviación) was created, but it was not until 15 August 1947 when the Military Paratrooper School (Escuela Militar de Paracaidistas) was founded, at the Alcantarilla AB. They jumped for the first time on 23 January 1948.

As a result of the reorganisation of the EdA in 1952, the unit was renamed First Paratrooper Squadron (Primer Escuadrón de Paracaidistas). It settled at Alcalá de Henares AB (Madrid) until it was ordered to participate in Ifni operations between 1957 and 1958, with combat launches and clean-up and security operations in the area. On 31 July 1958, it returned to its quartering in Alcalá de Henares, where it remained until its dissolution in 1965.

The Paratrooper Sappers Squadron was created by the Order of 9 September 1965, which dissolved the Paratrooper Squadron. This new unit was placed in the Military Paratrooper School 'Méndez Parada'.

The Squadron was made operationally dependent on the Tactical Aviation Headquarters, and administratively and logistically dependent on the School. From 11 March 1966, this operational dependence was exercised through the Colonel Director of the School.

On 18 November 1971, the squadron was ordered to join the Tactical Aviation Headquarters as another air unit. On 31 March 1973, it was assigned to El Copero Quartering, close to the Tablada Military Aerodrome.

On 26 April 1974, the unit returned to the Military Paratrooper School, where it remains to this day.

It is a Company type unit, with a high degree of specialisation. It is the only special forces unit in the EdA.

ABA, Academia Básica del Aire (Basic Air Academy)

Operationally depends on: MAPER (Personnel Command)
Organisationally depends on: MAGEN (General Air Command)

Military zone of Santiago de Compostela Airport. Seen is a C.101 Aviojet, E.25-51, 74-07. (Author's Photo)

General view of the 'Museo del Aire' at Cuatro Vientos AB, on the environs of Madrid. (Author's Photo)

Before the end of the Spanish Civil War, the School of Specialists was created at Málaga AB, which included all the specialities that the EdA had been teaching at various air bases up to that time.

In 1950, the School of Specialists moved to León, a school that was strengthened in 1953 because of the defence agreements between the United States and Spain. From then on, and with successive changes, the School of Specialists for non-commissioned officers includes:

- Basic Scale of the General Corps.
- Basic Scale of the Specialist Corps: Electronics Mechanics, Armourers, Radiotelegraphers, Photography and Cartography, Transmission Mechanics, Automobile Mechanics.
- Auxiliary Corps of Air Military Offices.

ETAC, EUROPEAN TACTICAL AIRLIFT CENTRE

The ETAC was created on 18 June 2017 at Zaragoza AB.

Tactical Airlift is a critical element in military operations that enables deployment.

The movement and support of air, naval and ground forces is essential in any emergency situation.

This unit is the successor, in the framework of NATO and the EU, of the 'European Air Transport' (2011) and the 'European Air Transport Training Course' (2012).

Zaragoza AB is the permanent headquarters where 11 countries work together.

The EADA is the unit in charge of monitoring the EVAs. (Author's Photo)

A member of the EADA with the high altitude launching equipment. (Author's Photo)

Specialised personnel from 15th Wing of Zaragoza handling an F-404 engine from an F-18 Hornet. (Author's Photo)

Between 1939 and 1952, it only served the then Eastern Air Region (Región Aérea de Levante) and was responsible for the repair of more than 30 aircraft models.

Between 1953 and 1968, it specialised in light aviation and in the 1970s, it moved on to serve the middle class aviation.

From the 1980s, it was commissioned to overhaul the Mirage F.1, CASA C.101 Aviocar, McDonnell Douglas F-18 Hornet, Eurofighter, Canadair CL-215 and Bombardier CL-415.

EADA, Escuadrón de Apoyo al Despliegue Aéreo, and SEADA, Segundo Escuadrón de Apoyo al Despliegue Aéreo (Air Deployment Support Squadron and Second Air Deployment Support Squadron)

Operationally depends on: MACOM (Combat Air Command)
Organisationally depends on: MAGEN (General Air Command)
The EADA is one of the EdA's most operational units, and possibly the most versatile.

Its mission is to support the defence, deployment, and maintenance of the operating conditions of air units.

It is a unique unit, not only because of its mission, but also because of the paratrooper status of its personnel and the plurality of its equipment.

It was created at Zaragoza AB on 12 January 1994, from the Military Air Transport Support Squadron (Escuadrilla de Apoyo al Transporte Aéreo Militar, EATAM), with the aim of supporting the Ícaro Detachment within the framework of the international operation 'Deny Flight' for the former Yugoslavia.

The SEADA was created at Morón AB in 2007.

MAESAL, Maestranza Aérea de Albacete (Albacete Air Workshops)

Operationally depends on: MALOG (Logistic Support Command)
Organisationally depends on: MAGEN (General Air Command)

After the creation of the EdA in October 1939, in November of the same year, it established the Material Services Section to be able to satisfy the maintenance and supply of all elements (aircraft and others).

At that time, in addition to the Maestranza of Albacete, there were seven others, each with their own speciality. The one in Albacete was moved to where it is today.

CLAEX, Centro Logístico de Armamento y Experimentación (Weapons and Experimentation Logistics Centre)

Operationally depends on: MALOG (Logistic Support Command)
Organisationally depends on: MAGEN (General Air Command)

Since 1991, CLAEX has inherited the functions carried out by the unit which, over the course of several years, had different names: Flight Test Group (1946), 406th Squadron (1968), 44th Air Force Group (1984) and 54th Wing (1986).

CLAEX is, together with the Air Workshops, the catalyst/multiplier for the development of the national aeronautical industries of applied engineering of high technological level.

Over the past 30 years (1991–2021), CLAEX has worked with all the weapon systems that have been incorporated into the F-18s and Eurofighters.

In recent years, it has created the Electronic Warfare Operational Support Squadron (Escuadrón de Apoyo Operativo a la Guerra Electrónica, ESAOGEL).

It has also been involved in the modernisation of aircraft such as: CASA-Northrop F-5, CASA C.295 and all transport and rescue helicopters. In the immediate future, this work will continue with the modern Eurofighter.

In terms of armaments, CLAEX has worked on the integration of the Litering and Reccelite pods, and the Iris-T, Taurus, and Meteor missiles.

An important chapter is the professional training in the speciality of armourers and the deactivation of explosives.

Its headquarters are located at Torrejón AB.

MDOP, Mando de Operaciones (Operations Command)

This institution is responsible for the planning, management and monitoring of surveillance, security, control, and air police operations in air spaces of sovereignty, responsibility, and national interest.

It plans and conducts multinational surveillance, control, security, and air police operations in and from, airspace when Spain assumes leadership and is determined by the Chief of Defence Staff.

MACAN, Mando Aéreo de Canarias (Canary Islands Air Command)

Operationally depends on: JEMA (Chief of the Air Staff)
Organisationally depends on: JEMA (Chief of the Air Staff)

This is the territorialised organisation of the EdA's General Air Command whose mission is the maintenance and preparation of the air units located in the Canary Islands, as well as to prepare the chiefs and officers.

It consists of a Headquarters under a Chief of Staff, to which the Gando AB with the 46th Wing and its 462nd Squadron, the Alert and Control Group (GRUALERCON) and SAR 82nd Group depend. Its force is based on the 46th Wing's F-18 Hornets.

MACOM, Mando Aéreo de Combate (Combat Air Command)

Operationally depends on: JEMA (Chief of the Air Staff)
Organisationally depends on: JEMA (Chief of the Air Staff)

This is the command unit of the EdA, resonsible for supporting and preparing air combat units, and providing the command and control system for offensive activities. Created on 3 May 1978 from the Air Defence Command (Mendo de la Defense Aérea) and home-based at Torrejón AB, it is operating in conjunction with MACO.

Its elements are:

11th Wing
12th Wing
14th Wing
31st Wing
35th Wing
Command and Control System Headquarters (Jefatura del Sistema de Mando y Control, JSMC)
Air Mobility Headquarters (JMOVA)
47th Air Force Mixed Group
Paratrooper Sappers Squadron (EZAPAC)
Aero Evacuation Medical Unit (UMAER, Torrejón)
Air Deployment Support Squadron (EADA)
Second Air Deployment Support Squadron (SEADA)
Aerial Medical Unit for Deployment Support (UMAAD-Madrid)
Aerial Medical Unit for Deployment Support (UMAAD-Zaragoza)

AGA, Academia General del Aire (General Air Academy)

Operationally depends on: MAGEN (General Air Command)
Organisationally depends on: MAPER (Personnel Command)

The General Air Academy is a military unit designed to train future officers of the EdA and is located at San Javier AB (Murcia).

It also houses a centre attached to the Polytechnic University of Cartagena.

Its origin is found in 1927. Before the Spanish Civil War, it was located in the now disappeared Alcalá de Henares AB (Madrid).

In 1943, the General Air Academy was moved to the San Javier AB.

The AGA is also the headquarters of the 'Patrulla Águila', the acrobatic group of the EdA.

Aircraft:
CASA C.101 Aviojet
ENAER T-35C Pillán

CESAEROB, Centro de Sistemas Aeroespaciales de Observación (Centre for Aerospace Observation Systems)

Operationally depends on: JEMAD (Chief of Defence Staff)
Organisationally depends on: MAGEN (General Air Command)

This department, created in 1994 on the occasion of Spain's participation in the HELIOS observation satellite programme, consists of:

The HELIOS España Main Centre, located at Torrejón AB, whose mission is the daily programming of Spanish missions, the processing and exploitation of satellite images, and the consultation of archives.

The Spanish Image Reception Centre (Centro de Recepción de Imágenes Español, CRIE), located in the Maspalomas Space Station (Gran Canaria), whose mission is to receive the images captured by the satellites. It is currently called the Aerospace Sensor Centre (Centro de Sensores Aeroespaciales, CRESAER)

GRUEMA, Grupo Escuelas de Matacán (Matacán Schools Group)

Operationally depends on: MAPER (Personnel Command)
Organisationally depends on: MAGEN (General Air Command)

Since 1939, Matacán AB has been home to the School of Flight Without Visibility (since 1939), the Higher School of Flight (since 1946), the School of Navigation and the School of Radio (since 1954) and the School of Air Traffic Control (since 1955).

In 1963, all these schools were unified under the name of *Grupo Escuelas de Matacán*. The GRUEMA's mission is to educate and train personnel in the tactics and techniques of the different areas of Military Air Transport, Air Traffic Control Techniques and to teach the courses that lead to obtaining the aeronautical qualification of Unmanned Aerial Systems Operator (UAV), as well as the training of personnel with the specialty of Fighting and Strike without destination in Air Force units.

GRUMOCA, Grupo Móvil de Control Aéreo (Mobile Air Control Group)

Operationally depends on: MACOM (Combat Air Command), through the JSMC (Command and Control System Headquarters)
Organisationally depends on: MACOM (Combat Air Command), through the JSMC (Command and Control System Headquarters)

This is the successor to the Tactical Aviation Control (created in 1966). In 1995 the unit was redefined with new tasks and renamed the Mobile Air Control Group.

GRUALERCON, Grupo de Alerta y Control (Alert and Control Group)

Operationally depends on: JSMC (Command and Control System Headquarters)
Organisationally depends on: JSMC (Command and Control System Headquarters)

It was created at Gando AB in 1987 to provide surveillance of the airspace assigned to it.

GRUALERCON is part of the NATO Integrated Air and Missile Defence System (NATINAMDS), depending on the Torrejón Combined Air Operations Centre (CAOC-TJ)

CN-ETAC, Centro Nacional-European Tactical Airlift Centre

Operationally depends on: MACOM (Combat Air Command)
Organisationally depends on: MACOM (Combat Air Command)

This is the National Component of the European Tactical Airlift Centre.

Based in the Zaragoza AB, it is the place from which the courses and exercises requested by the nations are planned and conducted. They also interact with other departments to advance training and interoperability issues.

GRUNOMAC, Grupo Norte de Mando y Control (Northern Command and Control Group)

Operationally depends on: MACOM (Combat Air Command), through the JSMC (Command and Control System Headquarters)
Organisationally depends on: MACOM (Combat Air Command), through the JSMC (Command and Control System Headquarters)

Its mission is the surveillance and control of airspace, and the exercise of tactical control with the means assigned to it for the execution of air police missions and any other entrusted to it, operating for this purpose in accordance with the provisions of the Concept of Operations of the Air Command and Control System.

It collaborates in the training and instruction of combat air units and supports the auxiliary forces in combat. Acts with other national and foreign armies, as required by the MACOM.

Escuela Militar de Paracaidismo 'Méndez Parada' (Military Parachuting School 'Méndez Parada')

Operationally depends on: MAPER (Personnel Command)
Organisationally depends on: MAGEN (General Air Command)

Its function is to train the paratrooper soldiers, physically and technically, so that they can serve in the various special units of this type.

They also train the personnel of the Army and Navy, the State Security Forces and Corps (Fuerzas y Cuerpos de la Seguridad del Estado, FCSE) and foreign police officers who are members of NATO.

UAS, Escuela de Sistemas de Vuelo no Tripulado (Unmanned Flight Systems School)

Operationally depends on: MAPER (Personnel Command)
Organisationally depends on: MAGEN (General Air Command)

It is located at Matacán AB, home to the Matacán Schools Group, and was created in June 2012.

MAESMA, Maestranza Aérea de Madrid (Madrid Air Workshops)

Operationally depends on: MAGEN (General Air Command)
Organisationally depends on: MALOG (Logistic Support Command)

Located in the Cuatro Vientos AB, it was created in 1940.

Its mission is to maintain the EdA's aircraft and helicopters.

CECAF, Centro Cartográfico y Fotográfico (Cartographic and Photographic Centre)

Operationally depends on: MAGEN (General Air Command)
Organisationally depends on: MAGEN (General Air Command)

It is in charge of the preparation and edition of photography and aerial cartography used by the EdA, in addition to many other functions related to the printing of manuals, documents and other similar items. This unit was created in 1979.
Aircraft:

Beechcraft C.90 King Air (U.22)
CASA CN.235 (T.19 y D.4)
Cessna Citation V (TR.20)

MDOA, Mando de Defensa y Operaciones Aéreas (Defence and Air Operations Command)

Operationally depends on: JEMAD (Chief of Defence Staff)
Organisationally depends on: JEMAD (Chief of Defence Staff)

It is the auxiliary unit of the JEMAD, integrated into the organisational structure of the Defence Staff as a permanent operational organisation.

Its objective is to plan, conduct and control military operations of surveillance, security, control, and air police of the air spaces of sovereignty, responsibility, and interest for Spain.

The Chief of the MACOM is also the Chief of the MDOA.

ESAOGEL, Escuadrón de Apoyo Operativo a la Guerra Electrónica (Electronic Warfare Operational Support Squadron)

Operationally depends on: MALOG (Logistic Support Command)
Organisationally depends on: MAGEN (General Air Command)

At the beginning of the 1990s, the need arose to create a department, eminently technical, in charge of centralising the management, research and advice on electronic warfare for all the EdA's air systems. This unit was set up as part of CLAEX.

In 1994 it began to operate, and in 1996 the unit carried out its first actions, which allowed it to accumulate practical experience. In this way, ESAOGEL became fully operational, which allowed it to be present in the 'Red Flag' exercise in the United States and the '*Nube Gris*' exercise in Albacete during those years.

This unit is deployed at Torrejón AB.

GRUCEMAC, Grupo Central de Mando y Control (Central Command and Control Group)

Operationally depends on: MACOM (Combat Air Command) through the JSMC (Command and Control System Headquarters)

The Getafe AB has workshops for the maintenance of its aircraft. (Author's Photo)

Organisationally depends on: MACOM (Combat Air Command) through the JSMC (Command and Control System Headquarters)

It originated in the former Combat Operations Centre (Centro de Operaciones de Combate, COC) and was created in 1992. It has staff, equipment, and facilities.

The GRUCEMAC, as the Main Centre of Air Defence, carries out the surveillance and control of the air space and the tactical control of the means assigned to it for the execution of air police missions and any other assigned to it.

CAOC-TJ, Combined Air Operations Centre-Torrejón

It is based at Torrejón AB.

It is an international unit within NATO's command structure whose mission is to plan, direct, coordinate, supervise, analyse, and report on air defence operations, following the guidelines of NATO's air component in its area of responsibility.

It originates from the 2010 Lisbon Summit, where a new strategic concept was implemented.

It currently has two CAOCs, one located in the German town of Uedem, responsible for the northern flank, and the CAOC-TJ, responsible for the southern flank, both under the command of a single Air Force Headquarters located at Ramstein AB (Germany).

MAGEN, Mando Aéreo General (General Air Command)

Operationally depends on: JEMA (Chief of the Air Staff)
Organisationally depends on: JEMA (Chief of the Air Staff)

The mission of the MAGEN is to provide the precise preparation of the auxiliary units of the EdA (except the training and flight test units), to facilitate the deployment of the units under its command in support of NBC defence activities, and to provide firefighting means, fuel, automotive, land-based weapons, et cetera.

SAR, Servicio Aéreo de Rescate (Rescue Air Service)

The SAR, which was born in Spain in the 1950s, has the mission of searching and rescuing, both at sea and on land, those EdA's pilots and any civilians who are in danger.

Born with a marked maritime vocation, its origins are to be found in the then Pollensa AB (Mallorca) and its area of activity, the Western Mediterranean. Over the years, the Service was also equipped with the necessary helicopters that would allow it to operate in mountainous areas that were difficult to access.

Since that time, it has carried out thousands of actions in every environment imaginable.

The SAR is organised on the basis of three flight squadrons which are distributed as follows: 801st Squadron based in the military area of Palma de Mallorca Airport (Son San Juan AB), from which it covers the whole of the Western Mediterranean; 82nd Group based at Gando AB (Las Palmas de Gran Canaria, Canary Islands) and covering one million square kilometres of the Atlantic Ocean; and 803rd Squadron based at Cuatro Vientos (Madrid), from which it reaches the whole of the Iberian Peninsula, as well as the Cantabrian Sea off the Spanish coast.

RCC-Madrid – RCC-Palma de Mallorca – RCC-Las Palmas de Gran Canaria (RCC: Rescue Coordination Centre):

These are the centres used for the coordination of SAR activities in a specific area. When an RCC is operating only in a maritime search and rescue area it is called MRCC (Maritime Rescue Coordination Centre)

CLOIN, Centro Logístico de Intendencia (Logistical Supply Centre)

Operationally depends on: MALOG (Logistic Support Command)
Organisationally depends on: MAGEN (General Air Command)

Until 1993, there were various service units, centres and organisations related to logistics in the EdA. The creation of CLOIN is the result of grouping them, in a logical way, considering their specific function.

In addition to the supply, it also carries out quality control of the materials it distributes. Its headquarters are located at the Torrejón AB.

CLOMA, Centro Logístico de Material de Apoyo (Logistics Centre for Support Material)

Operationally depends on: MALOG (Logistic Support Command)
Organisationally depends on: MAGEN (General Air Command)

It was created in 1993 to group together the following three units that had been dispersed for years: the Automobile Logistics Group (Grupo Logístico de Automóviles), the Chemical Defence and Firefighting Logistics Squadron (Escuadrón Logístico de Defensa Química y Contraincendios) and the Fuel Logistics Squadron (Escuadrón Logístico de Combustibles). The CLOMA is located at Getafe AB.

CLOTRA, Centro Logístico de Transmisiones (Transmission Logistic Centre)

Operationally depends on: MALOG (Logistic Support Command)
Organisationally depends on: MAGEN (General Air Command)

It was created in 1939 as the Aviation Transmission Regiment.

From that moment on, it evolved and changed its location. It is currently located in the Getafe Quartering (ACAR Getafe).

Its mission is the training of transmission-related weapon systems.

ESTAER, Escuela de Técnicas Aeronáuticas (School of Aeronautical Techniques)

Operationally depends on: MAPER (Personnel Command)
Organisationally depends on: MAGEN (General Air Command)

It was constituted in 1994 under the name of Centre of Aeronautical Techniques (Centro de Técnicas Aeronáuticas, CETAER).

The EADA has all types of light armament for the defence of bases and other facilities. (Author's Photo)

It provides specific technical and practical training for aspiring career and non-career military students in the EdA's Corps of Engineers.

EMACOT, Escuela de Técnicas de Mando, Control y Telecomunicaciones (School of Command, Control and Telecommunications Techniques)

Operationally depends on: MAPER (Personnel Command)
Organisationally depends on: MAGEN (General Air Command)

Its mission is to provide training and further education to train EdA's personnel in Command and Control, and Transmissions and Electronics.

TEACHING CENTRES

CENTRE	LOCATION
Academia Básica del Aire (ABA)	León Military Aerodrome
Escuela de Especialidades	Torrejón AB
Escuela de Automovilismo	ACAR Getafe
Academia General del Aire (AGA)	San Javier AB
Escuela Militar de Paracaidismo	Alcantarilla AB
Centro de Adiestramiento de Seguridad y Defensa (CASYD)	ACAR Los Alcázares
Grupo Escuelas de Matacán (GRUEMA)	Salamanca AB
Escuela de Apoyo (ESAPO)	ACAR Tablada
Escuela de Técnicas de Seguridad, Defensa y Apoyo (ETESDA)	Zaragoza AB
Escuela de Técnicas Aeronáuticas (ESTAER)	Torrejón AB
Escuela de Técnicas de Mando, Control y Telecomunicaciones (EMACOT)	Cuatro Vientos AB
Centro Técnico de Apoyo de Hostelería, Ofimática, Abastecimiento y Oficios de Mantenimiento	Zaragoza AB
Escuadrilla Logística de Chinchilla	Torrejón AB
Escuela de Pilotos de Helicóptero	Armilla AB

Its students come from the Schools for Officers, Non-Commissioned Officers and Professional Military Troops, in the specialities of Command and Control and Transmission Mechanics and Electronics.

Escuela de Pilotos de Helicóptero (Helicopter Pilot School)

It is deployed at Armilla AB near Granada. The main function of the School is teaching related to the official Spanish State helicopters. Courses are also given in Helicopter Maintenance Mechanics, for troop personnel of the EdA and the Guardia Civil.

This school is attended by future pilots of this type of

aircraft from the EdA, the Guardia Civil, the National Police and the Customs Surveillance Service.

ETESDA, Escuela de Técnicas de Seguridad, Defensa y Apoyo (School of Security, Defence and Support Techniques)

Operationally depends on: MAPER (Personnel Command)
Organisationally depends on: MAGEN (General Air Command)

Its mission is to generate training and improvement of the EdA's personnel in the areas of security, defence, police dog training, VIP protection, and hospitality, as well as the military training of the troop. It also trains EdA's volunteer reservists in the specialities to be determined.

GRUTRA, Grupo de Transmisiones (Transmission Group)

Operationally depends on: JSTCIS (Head of Technical Services and Information and Telecommunications Systems)
Organisationally depends on: MAGEN (General Air Command)

It is one of the oldest units of the EdA. Created in 1939 as the Aviation Transmission Regiment (Regimiento de Transmisiones para la Aviación), as an organic unit of the Army.

The Unit is deployed at Getafe AB.

MAESE, Maestranza Aérea de Sevilla, (Seville Air Workshops)

Operationally depends on: MALOG (Logistic Support Command)
Organisationally depends on: MAGEN (General Air Command)

Between 1921 and 1925 it was called 'Talleres del Aeródromo de Tablada' (Tablada Aerodrome Workshops) and between 1926 and 1940 it was known as 'Parque Regional del Sur' (Southern Regional Park).

Today its function is to maintain the weapons systems and to issue and renew the airworthiness certificates of the Rota aircraft (of the Spanish Navy Air Force), and of the Cadets Corps Foundation 'Vara de Rey', in Corral de Ayllón (Segovia).

UMAAD-Madrid, Unidad Médica Aérea de Apoyo al Despliegue, and UMAAD-Zaragoza, Unidad Médica Aérea de Apoyo al Despliegue (Aerial Medical Deployment Support Units, Madrid, and Zaragoza)

Operationally depends on: MACOM (Combat Air Command)
Organisationally depends on: MAGEN (General Air Command)

Deployed at the Torrejón and Zaragoza Air Bases, they were created in 2003, with the mission of providing the health care (preventive, specialised and surgical) required by military forces deployed in real actions or exercises. They are also responsible for evacuations carried out by an Aero Evacuation Medical Unit (UMAER).

UMAER, Unidad Médica de Aeroevacuación (Aero Evacuation Medical Unit)

Operationally depends on: MACOM (Combat Air Command)
Organisationally depends on: MAGEN (General Air Command)

The Aero Evacuation and Immediate Intervention Medical Unit is a unique unit that is difficult to compare with any other of similar dimensions, due to its tasks, its particular idiosyncrasy, and its specific operating materials.

Its main missions are: air evacuation, cooperation in search and rescue missions, support to medical units supporting the deployment of the EdA, and immediate medical interventions.

MAPER, Mando de Personal (Personnel Command)

It depends directly on the JEMA.

This command manages and controls the human resources of the EdA's staff. It also manages the list of military posts and provides training and further education for the EdA's military personnel.

CIMA, Centro de Instrucción de Medicina Aeroespacial (Centre for Aerospace Medicine Education)

It is the military health section specialised in aeronautical medicine, which depends on the MAPER.

It is based at Torrejón AB.

JMOVA, Jefatura de Movilidad Aérea (Air Mobility Headquarters)

Operationally depends on: MACOM (Combat Air Command)
Organisationally depends on: MACOM (Combat Air Command)

Created in 2004, it has the personnel of the now extinct '*Mando Aéreo de Levante*' (Levante Air Command) supported by the personnel and technical means of the Zaragoza AB.

It consists of a Head Office, General Secretariat, Air Transport Secretariat, Deployment Support Secretariat, and an Operational Logistics Resources Section.

It is deployed at the Zaragoza AB.

JSMC, Jefatura del Sistema de Mando y Control (Command and Control System Headquarters)

Operationally depends on: MACOM (Combat Air Command)
Organisationally depends on: MACOM (Combat Air Command)

This Head Office was created in 1999 as an operational and technical organisation to centralise the activities of the Air Command and Control structured on the basis of the GRUCEMAC, GRUNOMAC, GRUALERCON and GRUMOCA, as well as the Operational Air Circulation Group (GRUCAO).

It has the direction, coordination, and evaluation of all the functions of the Command and Control System.

COVE, Centro de Operaciones de Vigilancia Espacial, (Space Surveillance Operations Centre)

It depends on the Command and Control System Headquarters (Jefatura del Sistema de Mando y Control) of MACOM, and entered into service in December 2019 at Torrejón AB

Its task is to predict the space risks that may affect Spain and thus to reduce their effects as much as possible.

Its mission is space surveillance and the provision of services in support of other operations carried out by the armed forces.

Its creation, according to the EdA, represents a fundamental step in the process of acquiring the space surveillance capability begun in 2015, when Spain assumed operational control of the European Space Agency's (ESA) space surveillance radar located in Santorcaz (Madrid).

The system is based on the S3-TRS radar located in Morón de la Frontera. In addition, the COVE has control of the Santorcaz radio station, for tracking and detection in low orbits (200 to 2,000 kilometres) and has access to the data provided by the telescopes in Puertollano (Ciudad Real), Montsec (Lleida), Teide (Tenerife) and San Fernando (Cádiz).

ACGEA, Agrupación del Cuartel General del Ejército del Aire (EdA's Headquarters Group)

Operationally depends on: JEMA (Chief of the Air Staff)
Organisationally depends on: JEMA (Chief of the Air Staff)

It is the Security Service of the General Headquarters of the EdA, and the 'Casa del Suboficial' (House of the Non-commissioned officer).

SHYCEA, Servicio Histórico y Cultural del Ejército del Aire (EdA's Historical and Cultural Service)

Operationally depends on: JEMA (Chief of the Air Staff)
Organisationally depends on: JEMA (Chief of the Air Staff)

Created in 1972, under the name 'General and Historical', it is currently the institution responsible for the cataloguing, protection, conservation, research, and diffusion of the historical and cultural heritage of the EdA.

It brings together the Institute of Aeronautical History and Culture, and the Museum of Aeronautics and Astronautics of the EdA.

The archive is located in the castle of Villaviciosa de Odón and the museum in the Cuatro Vientos Air Base, both in the province of Madrid.

Museo de Aeronáutica y Astronáutica, 'Museo del Aire' (Museum of Aeronautics and Astronautics)

Once the EdA was created (1939) after the Spanish Civil War, the idea was to have a museum where, as a legacy to the future, aircraft and other objects reflecting the history and evolution of Spanish aviation would be exhibited. The collection that was gathered was installed in the basement of the then Ministry of the Air, now the EdA's Headquarters (Plaza de la Moncloa, Madrid).

The Museum was legally created on 16 June 1966, and the collection was moved to the place it now occupies at Cuatro Vientos AB, although on a provisional basis, since for more than 9 years it acted more as a warehouse than a museum. Furthermore, the refurbishment work did not begin until 1979 and its doors were not opened until May 1981.

9
PEACEKEEPING AND HUMANITARIAN AID MISSIONS OF EJÉRCITO DEL AIRE

In the past, the EdA has contributed, by decision of the Spanish government or at the request of international organisations such as the United Nations (UN), the North Atlantic Treaty Organization (NATO), the European Union (EU) or the Western European Union (WEU), by sending humanitarian aid or by intervening in peace missions due to the outbreak of war.

Three types of mission have been carried out: sending combat aircraft (F-18 Hornet and Eurofighter Typhoon); sending transport aircraft to deliver humanitarian aid to very distant locations (C-130 Hercules, CN.235, DC-8, C.295 and C.212 Aviocar); and dispatching military personnel of the EdA to take part in peace, control, and interposition missions.

One of the longest and most distant missions has been in the Indian Ocean, 'Operation Atalanta', using our P-3B Orion, whose task was to patrol the waters off the east coast of Africa.

Several thousand EdA men and women have contributed to the huge task of transporting and distributing the material aid required. Thousands and thousands of flight hours have been accumulated over several decades of that activity.

Fortunately, the cost in terms of human lives and aircraft has been minimal, as there is only one air accident to record, which occurred on 2 January 1987, when a C.212 Aviocar fell into the sea in Equatorial Guinea with the loss of its crew and the people it was carrying, 18 in all.

Missions in the Framework of the UN, NATO, WEU and EU, and those Commissioned by the Government of Spain

Perhaps the oldest of these was the dispatch of EdA's transport aircraft with humanitarian aid to mitigate the damage caused by the earthquake in Agadir (Morocco) in February 1960.

- Earthquake in Algeria (1980)
- Cameroon, volcanic eruption (1986)
- Floods in Bolivia (1986)
- El Salvador Earthquake (1986)
- Mexico Earthquake (1986)
- Locust plague in North Africa (1987–1988)
- UNAVEM I and UNAVEM II in Angola (1989–1991 and 1991–1995)
- ONUCA and ONUSAL in Central America (1989 and 1990–1991)
- Mauritania and Senegal, race riots (1989)
- ONUVEH, Haiti (1990)
- Evacuations in Liberia (1990)
- Iran Earthquake (1990)
- First Persian Gulf War (1990–1991)
- Evacuations in Zaire (1991)
- Iraq, 'Operation Provide Comfort' (1991)
- Somalia Civil War (1991–1993)
- ONUMOZ mission in Mozambique (1992–1994)

Conflict in former Yugoslavia (UNPROFOR, WEU, EU, 1992–1996)
UNAMIR Mission in Rwanda (1994–1995)
Earthquake and Tsunami in Indonesia (2004)
EU-NAVFOR 'Operation Atalanta' in Somali waters (2008–2020)
Forest Fires in Israel (2010)
Chile Earthquake (2010)
'Operation Unified Protector' (Libya, 2011)
NATO air police missions in the Baltic (2017–2020)
Police air mission in Romania. Paznic Detachment (2021)
Mission accomplished: evacuation of Kabul

Current Actions Abroad

In recent years, the EdA has participated, and continues to do so, in different scenarios around the world, providing support to different international missions, in accordance with the mandates of the UN, NATO and other organisations. Missions listed as follows:

During February and March 2021, the EdA organised for the first time the Paznic Detachment to operate from Romania, thus contributing to the reinforcement of the Atlantic Alliance's air police in the strategic Black Sea region. This detachment will be activated at NATO's request in the future.

Baltic Air Police (BAP) operation. The last intervention was carried out between March and September 2019 from Siauliai AB (Lithuania), by an Aero-Tactical Detachment made up of 13 military personnel and five F-18M Hornet from the 12th Wing. 484 departures and 740 flight hours were performed.

Destacamento Orion (Orion Detachment). Deployed in Djibouti, it is participating in the European Union's NAVFOR 'Operation Atalanta' Somalia to combat piracy and protect the ships of the UN World Food Programme (WFP). During 2019, the detachment consisted of a P-3B Orion and 60 military personnel who carried out 150 missions and nearly 1,000 hours of flying time.

Destacamento Marfil (Marfil Detachment), deployed since January 2013 in Dakar (Senegal), is made up of a C.130 Hercules and 52 military personnel. Its mission is to carry out transport operations within the framework of the French 'Barkhane' operation and the UN missions MINUSMA and MINUSCA, to stabilise the Republic of Mali and the whole of the Sahel. In 2019 they carried out 400 missions and 1,100 flight hours.

Within the framework of the Destacamento Mamba (Mamba Detachment), the EdA supported French operations in several countries of the Sahel. The Spanish presence is based on a C.295M stationed in Libreville (Gabon). During 2019, 450 missions and 1,100 flight hours were completed.

Since 2015, the Destacamento Aéreo 'Grappa' ('Grappa' Airborne Detachment) has been deployed to Sigonella AB (Sicily, Italy). A C.235 of the SAR and 40 military personnel are stationed there in Operation EUNAVFORMED Sophia, which aims to control illegal immigration into southern Europe. During 2019, 150 missions and 900 flight hours were carried out.

In Operation 'Iraq Support' (Operación 'Apoyo a Irak'), which began in 2015, the EdA maintains a permanent Tactical Attack Control Team (TACP). The team is made up of five military personnel from the Building Party Capabilities (BPC) based in Besmayah. In the same operation, since 2018, an air-surface integration team (SOALI, Special Operations Air Land Integration) made up of five military personnel belonging to the Special Operations Task

Spanish military personnel descending from a C-130 Hercules at Sarajevo Airport (Bosnia-Herzegovina). (Author's Photo)

Spanish aircraft at Mostar Airport (Bosnia-Herzegovina). (Author's Photo)

Group (SOTG), whose usual area of operations is Baghdad and its surroundings, has been deployed.

As part of the fight against international terrorism, the EdA is supporting NATO's 'Operation Sea Guardian', whose mission is to monitor maritime areas in the Western Mediterranean. It operates with a P-3C Orion and a C.235 from the VIGMA-SAR. In the course of 2019, it carried out 38 missions and 230 flight hours.

In connection with the European Border and Coast Guard Agency's (FRONTEX) 'Operation Indalo', the EdA began its participation in 2019 with an aircraft that completed 48 departures and 207 flight hours.

EdA's personnel and materials collaborated in various Cooperative Security tasks in Senegal and Mauritania, aimed at training and advising the armed forces of these countries.

The Headquarters of 'Operation Resolute Support', in Kabul, Afghanistan, has an EdA's liaison.

Finally, it should be mentioned that this armed force maintains its presence in various international organisations, including the Euro-NATO Joint Jet Pilot Training (ENJJPT); Joint Air Power Competence Centre (JAPCC); European Tactical Airlift Centre (ETAC); Air Component Euro Corps (ACEC); European Air Group (EAG); NATO Alliance Ground Surveillance (NATO-AGS); and NATO Eurofighter and Tornado Management (NETMA).

War Action in Bosnia-Herzegovina

On 8 March 1994, a C.212 Aviocar from the 37th Wing based at Villanubla on a UN operation 'Deny Flight' was attacked during an air transport mission between Zagreb and Split, Croatia, in the territory of the former Yugoslavia. The flight was undertaken for the transfer of NATO and UNPROFOR personnel, in addition to the transport of general cargo and ammunition. The crew's experience allowed for an emergency, single-engine landing. The Aviocar was apparently hit by anti-aircraft fire. NATO, which described the attack as a provocation, said it occurred at 12:35 local time when it was flying 55 kilometres east of the town of Rijeka, near the region known as Croatian Krajina, which was occupied by the Serbs at the time.

The C.212 Aviocar was hit in the left engine and suffered other damage to the fuselage and tail. At the time, the aircraft was flying at 7,000 feet (2,100 metres). Subsequent analysis showed that it received medium-calibre artillery fire, possibly 20 millimetres, or from the shock wave of a missile, possibly a SAM-7, which fortunately did not hit the aircraft, which would otherwise have been shot down. The flight had been proceeding normally up to the time of impact, as the weather conditions were excellent. The jolt and rumble from the impact was brutal, with a heavy jolt and thick whitish smoke penetrating the fuselage. The impact affected the tail of the aircraft, the tailgate, the left engine oil radiator and the interior, but no one was injured.

The pilot immediately stopped that engine to avoid the fire. The incident resulted in a loss of power and subsequent loss of altitude.

On the ground and after a detailed inspection, a fragment of an electronic component with an inscription identifying the device as a SAM-7B was found. A total of 149 impacts were detected, most of them centred on the shattered tail and severe damage to the vertical rudder. After 72 hours of work at Rijeka airport, the aircraft was recovered and put into flight.

31st Wing Hercules Involvement in 'Operation Unified Protector'

On 20 March 2011, two crews from the 31 Wing based in Zaragoza, were activated. Their mission was to deploy material and personnel from the 12th Wing and the 47th Group to the Decimomannu air base (Sardinia Island, Italy). On 18 March, the UN Security Council had authorised its member states to take the necessary measures to protect Libyan civilians and to establish a no-fly zone, a weapons embargo and a humanitarian corridor. A month later, a KC-130H (TK.10) from the 31st Wing and a Boeing 707-331B (TK.17) from the 47th Group were deployed there, tasked with in-flight refuelling of allied fighters operating in Libya to maintain the no-fly zone.

Initially, Allied missions in Libyan skies were night operations, so the EdA's Hercules flew four nights in a row, rested two, flew another two, and rested the next night. As the conflict progressed, operations became daylight operations. After more than six months of activity, the missions were spaced out and ended in October of the same year.

One night mission of the Spanish aircraft was to fly to the refuelling area off the Libyan coast, stay there for an hour and a half and return to Decimomannu at around 07:00 in the morning.

Forty-five day rotations were established in order to give all flight and ground personnel the necessary rest. At the end of the six months of activity, the Hercules flew 118 flights, providing one million litres of fuel to the different types of fighter aircraft involved in the mission.

BALTIC SEA AND BLACK SEA AIR DETACHMENTS

Baltic Sea

In the NATO framework, the EdA conducted two airborne police detachments in the Baltic Sea and the Black Sea. The first of these began its activity on an ad hoc basis in 2004, although it has been active on a regular basis since 2015, whilst the second was initiated in 2021 and will continue to be active in the future. These actions are conducted in combination with other NATO air forces.

The Vilkas Tactical Air Detachment carries out air policing missions in the Baltic from the Latvian Siauliau Air Base. These detachments are manned by F-18M personnel and aircraft from the 12th Wing at Torrejón. Between pilots and ground personnel, some 130 men and women from the unit itself, the EADA, UMAAD, DAE, GRUMOCA, CIJES, JSMC, GRUNOMAC, as well as personnel from the Torrejón Air Base itself, usually travel to the base. In the framework of the rotations, which last about three months, about 500 sorties are carried out, usually totalling about 750 flying hours.

During these detachments, the airspace of the Baltic States is monitored through actual missions, but also through training flights scheduled by the CAOC Uedem, involving aircraft from other NATO countries stationed in the area. The actual missions consist of intercepting Russian fighter and transport aircraft flying over the Baltic Sea and surrounding waters, and the skies of the three countries under surveillance.

Until mid-2021, the EdA has been conducting such missions seven times.

Black Sea

Following the events in Ukraine and Crimea in 2014, NATO activated an operation for the surveillance and control of the Black Sea. Spain and the EdA responded to the call by setting up a working group to lay the foundations for future collaboration.

At the time of closing this chapter, only one detachment has been carried out, which was deployed to the Romanian Mihail Kogalniceanu air base near the Black Sea coastal city of Constanta at the end of January 2021. Six Typhoon of the 11th Wing from the Morón Base were deployed there for three months, along with some 130 men and women from different units of the EdA. They are all part of the Paznic Tactical Air Detachment, which is tasked with surveillance and control of the wider Black Sea area. The aim is to contribute to deterrence in the area against Russian forceful actions. In doing so, NATO is demonstrating the collective commitment of Alliance members.

During this first mission, the Spanish aircraft flew 400 sorties during which real interdiction and training missions were carried out, the latter in collaboration with Romanian F-16s and MiG-21s and French Rafale aircraft. It should be noted that in March 2021, Spanish Typhoons intercepted two Russian Sukhoi Su-30 fighters flying without transponder signals and flight plan and without contact with the civilian Air Traffic Control Centre.

These missions will continue in the future, as a demonstration of the capabilities of the EdA and Spain's willingness to contribute to the defence of world peace.

Mission Accomplished: Evacuation of Kabul

Experience tells us that the success of a military operation depends to a large extent on good planning, the time needed to prepare for it and having the right equipment to carry it out. In the case in question, it can be said that these premises could not be entirely fulfilled since, given the circumstances that initially surrounded it, what prevailed was improvisation, although the experience accumulated over decades by the EdA contributed to the achievement of the objective.

It is well known that at the beginning of 2020, US President Donald Trump and representatives of the Taliban agreed that US troops would be gradually withdrawn from Afghanistan. NATO officials subsequently declared the termination of its 'Operation Resolute Support' by 21 September 2021. Both said they saw the Afghan government as capable of defending its regime on its own, provided it could rely on the war materiel that Western forces had stationed in the country.

Following this decision, Spain synchronised the departure of its troops with that of the other allies, a commitment it reached in May 2021, after almost 20 years of effort, and permanence.

After NATO's announcement, events were accelerating and the Taliban, anticipating what was to come, began a general offensive across the country to regain ground in order to strike the final blow before the deadline agreed the previous year. In early August 2021, international observers hypothesised that the Afghan regime would not be able to hold out beyond the autumn, not imagining that the Taliban would be able to attack Kabul, the Afghan capital, within days. These precipitous events say little for the intelligence services of the already few Western troops then present in the country.

The Operation gets Underway

When Western governments realised on 12 August that the situation in Kabul was becoming unsustainable, the Spanish Defence Staff convened the Operational Planning Group, in which the Operations Command (MOPS, *Mando de Operaciones*), the Joint Defence Staff (EMACON, *Estado Mayor Conjunto de la Defensa*) and the Army (ET, *Ejército de Tierra*) took part. The Air Force was represented by members of the Air Force Staff (EMA), Combat Air Command (MACOM) and Air Mobility Headquarters (JMOVA). This videoconference laid the groundwork for a NEO (Non-Combat Evacuation Operation), aimed at removing from Kabul, the Afghan civilian personnel who had been collaborating for the past 20 years with both the Spanish embassy in the capital and the Spanish troops stationed in that country.

The situation, which was addressed quickly and effectively, allowed the Spanish Armed Forces to participate in the 'Resolute Support' mission. At the same time, the activation of Operations Plan 00066 was authorised, directed by the Commander of the Command (OPLAN-CMOPS), which established the framework for a military operation in support of the Ministry of Foreign Affairs, European Union and Cooperation.

Operational command rests with CMOPS, which ordered the immediate preparation of the capabilities and readiness of the units to be involved in the operation.

On 14 and 15 August, as the situation worsened, the operation was accelerated in anticipation of evacuating 800 people from Kabul. The new Airbus A.400 Atlas were deemed capable of this task and could fly from Kabul airport to an intermediate airport or airbase where the rescued personnel could be transferred to civilian flights contracted by the Ministry of Defence.

The operation should have AMPT (Air Mobility Protection Team) integrated among the Atlas crews and on the contracted flights of Air Europa (the selected Spanish civilian company). Medical teams should also be available to assist military and civilian personnel to be evacuated, both on civilian and military flights. MEDEVAC (Medical Evacuation) capability should be provided by adapting one of the A.400 Atlas for this purpose.

For all these reasons, the 31st Wing, based in Zaragoza, was designated as the ideal unit to carry out the transport and evacuation. The EADA (Air Deployment Support Squadron), which would include the UTER (Unidad Terminal, Terminal Unit) and the UMAER (Air Evacuation Medical Unit), which has MEDEVAC capacity and personnel from the DISAN (Dirección de Sanidad, Directorate of Health) for medical support, would be among the capabilities provided by the EdA. The Army was to contribute elements of the Civil Military Cooperation Battalion (CIMIC), and medical personnel.

First Difficulties

In order to successfully accomplish the mission and due to the impossibility of refuelling the aircraft in Kabul, the aviation officials agreed to have an intermediate airport at which to refuel, rest and transfer the evacuated personnel to aircraft of a civilian company. After consultation, Al Minhad Air Base, close to Dubai airport in the United Arab Emirates (UAE), was chosen.

On 16 August, the first Airbus A.400 Atlas of the 31st Wing, TK.23-07, 31-27 (AME3165), left for the base to transport members of the EADA, CIMIC and medical personnel to establish a forward base. The following days, 17 and 18 August, T.23-10, 31-30 (medicalised) (AME3166) and T.23-04, 341-24 (AME3167) flew to this destination with the rest of the personnel initially planned.

On 18 August, the first entry into Kabul was made, where the EADA personnel who make up the UTER were deployed. On the return flight, 53 Afghans were taken out of the country.

Kabul airport became a chaotic place at this time, difficult to control and with runways frequently overrun. Despite the fact that the airport was under the command of US troops, incidents were extraordinarily frequent and many remember the attempts to storm the access gates, groups of men trying to board aircraft at every stage of manoeuvres and finally, a terrible attack in which more than 130 people perished, including many US military personnel.

Another aspect to be considered was that aircraft could only remain on the ground for 30 minutes and that take-offs had to be accompanied, as a precautionary measure, by countermeasure launches. Another element related to the transfer of refugees, who could only stay on Emirati territory just long enough to transfer from a military aircraft to a civilian one, which had to take off immediately. This was achieved thanks to the good understanding between the MOPS (Mando de Operaciones, Operations Command) and Air Europa personnel.

It is worth mentioning that the UME (Unidad Militar de Emergencias, Military Emergency Unit) had to move two 4X4 VW Amarok vehicles to Kabul, on loan from the UME Battalion in Zaragoza, as well as a large amount of material that had to be abandoned in Kabul at the end of the mission.

Another element that in an improvised way, had to be moved to Kabul, was an MOE team (Mando de Operaciones Especiales, Special Operations Command) with all its equipment, hypothetically destined to operate outside the perimeter of the Kabul airport. At their destination, they were integrated into the UTER.

Within a few days the situation at Kabul airport worsened. On the one hand, the increasing number of aircraft arriving to take out diplomats, embassy staff and Afghan personnel who had been in the service of the international troops, made it very dangerous for the aircraft to remain on the ground. The airport itself was under pressure and threatened with invasion and attack. On the other hand, the intelligence services were not acting in the desired manner and after the very serious attack on 26 August, the worst was feared. On the Spanish side, all diplomatic personnel were concentrated at the airport given the chaos that prevailed in the city.

The situation at the airport reached the point where the command took the decision that the members of the Army's CIMIC Battalion would remain at the UAE's Al Minhad Air Base, leaving only the members of the UTER and intelligence personnel deployed in Kabul.

Thanks to the important work of the Spanish members of the Combined Air Operations Centre (CAOC) in Qatar, from where air movements in this area of the Middle East were coordinated, very detailed intelligence reports were received at all times from the coalition. This enabled risk management decisions to be taken, operating at all times with acceptable but changing security information.

At that time, the two Spanish A.400 Atlas deployed to Afghanistan were able to carry out two daily rotations (Al Minhad-Kabul-Al Minhad) and occasionally up to three, despite the precariousness of the operations (with little support and major operational limitations).

A fourth aircraft, TK.23-02, 31-22 (AME3169), was dispatched on 19 August to replace T.21-10, 31-30, which had repatriated the first Afghans a few days earlier and was carrying vehicles and special operations equipment on its return trip to the area of operation.

At that time, three Atlas and one Air Europa aircraft were already deployed in Al Minhad.

The information obtained and the news from Kabul, prompted the general command of the international rescue operation to decide to bring forward the end of the evacuation to 31 August.

Meanwhile, on 25 August, the flight hour potential of the first Atlas dispatched, TK.23-07, 31-27 (AME3165), was used up and had to be relieved by a fifth aircraft, T.23-05, 31-25 (NAG16FN).

From a human and technical point of view, it is worth mentioning that for the Spanish crews it was a very strenuous activity. An hour before take-off, there was a general briefing with the entire crew, which was reinforced by an extra safety pilot, a third load supervisor, an EADA safety team, a medical team, and a team of two maintenance certifiers.

The established rule for each mission was that time and operations at Kabul airport had to be minimised. It was therefore decided to carry out all operations with the engines running, thus saving time and reducing the chances of technical failures in successive start-ups.

The number of evacuees was to be as large as possible but not to exceed the total capacity of the Air Europa aircraft. This is one of the reasons why refugees were transported seated on the floor with straps fastened to the floor, as on several flights, up to 150 passengers were loaded on board. The flight time between Kabul and Al Minhad, including an hour's stay at Kabul airport to embark refugees, was about eight hours. This time was increased by an additional four hours spent in Al Minhad unloading and completing formalities.

In summary, 1,891 Afghans, eight military personnel of other nationalities and 19 people from the Spanish Embassy in Kabul were evacuated, all of this was carried out in 17 Atlas flights.

The evacuation operation in Afghanistan was characterised by the fact that it took place in a highly changing scenario, where the rapid advance of the Taliban caused events to rush. This meant that the personnel of the Air Deployment Support Squadron (EADA) had to adapt to tasks initially assigned to the Spanish Embassy, which demonstrated the versatility of the Unit.

10

'APAGA Y VÁMONOS', 50 YEARS OF ACTIVITY OF THE 43RD AIR FORCE GROUP

At the end of 2021, the 43rd Air Forces Group of the EdA celebrated 50 years of activity with a positive balance.

Based in Torrejón de Ardoz (Madrid), the 43rd Group's main mission is to extinguish forest fires and its secondary mission is to support SAR missions. It is currently equipped with 14 Canadair CL-215 (UD.13) and four Canadair CL-415 (UD.14).

Various agreements and protocols signed since 1971 have regulated the conditions for the use by the Ministry of Defence of the amphibious aircraft historically purchased by the Ministry of Agriculture. An interdepartmental agreement currently governs the collaboration between the two ministries to meet the operational, maintenance and availability needs of these aircraft.

The 14 amphibious aircraft currently operated by the EdA depend operationally on the Army's Military Emergency Unit (UME) and functionally, on the Ministry of Ecological Transition and Demographic Challenge, which is the heir to the Ministry of Agriculture that purchased and used the Canadair.

This was the 3rd Air Group's first CL-215. First leased, finally purchased in 1971. Note that it has no radar and was powered by 18-cylinder Pratt & Whitney engines. It was, with codes UD.13-1, 43-01, in service until 1991. It is currently housed at the Cuatro Vientos Air Museum. (Author's Photo)

Aerial view of UD.13-10, 43-10 during a water intake demonstration in a reservoir. This model was already equipped with weather radar. (Author's Photo)

The unit's coat of arms bears the motto 'Apaga y Vámonos', a motto under which thirty-four CL-1215/CL-415 have operated over the years, moving them during the period when the most forest fires tend to occur (15 June/30 September) from Torrejón AB to the detachments located at the bases and airports of Santiago de Compostela, Zaragoza, Albacete, Málaga, Palma de Mallorca, Salamanca, Talavera de la Reina, and other destinations from which they carry out their missions. (*'Apaga y Vámonos'* – Switch off and let`s go – is a colloquial Spanish phrase used when something is finished).

Their mission is one of the riskiest jobs in the aeronautical environment: flying at low altitude, in the midst of smoke that reduces visibility, with strong turbulence that makes flying difficult, sometimes in the midst of a concentration of aerial resources and in rugged geographical areas – all of which requires constant training.

At its base in Torrejón is seen UD.13-11, 431-01. Once retired in 1995, this aircraft was sold to a civilian company that registered it EC-GBR. (Author's Photo)

The current 43rd Air Forces Group is the heir to 404th Squadron, later 803rd Squadron, units always based at Torrejón.

During its 50 years of activity, the Canadair has flown 185,000 flight hours, of which 86,000 have been real missions against forest fires, both in Spain and abroad, in support of other nations such as Portugal, Morocco, Italy, Greece, Bulgaria and Israel, the latter being the most geographically distant of the countries that have requested its collaboration.

The 34 aircraft that have passed through this unit have had exceptional flight and technical support personnel. Of all of them, mention must be made of the 15 crew members who died in action.

It must not be forgotten, the work carried out during those five decades by the Albacete Air Workshops, the technical head of that fleet and without which, it would have been impossible to carry out this important aeronautical work.

11 EJÉRCITO DEL AIRE STAFF

Making an exact calculation of the staff (men and women) that currently make up the EdA is extremely difficult, given that, among other things, there is a data protection law that prevents access to official lists.

The numbers that follow should be taken only as an approximation.

Pilots preparing for a training flight in a two-seater F-18 Hornet. (Author's Photo)

After a training flight. (Author's Photo)

SAR crew after a demonstration flight. (Author's Photo)

Preparing for a flight in an F-18 Hornet, a pilot with one of the unit's mechanics as his assistant. (Author's Photo)

It must be considered, as mentioned at the end, that among the most prepared and experienced personnel (generals, chiefs, and officers) there is an important mobility, given that the private business sector (airlines, high technology industries and similar) has a great appreciation for the EdA personnel, to whom they offer better professional expectations and better economic remunerations. This phenomenon has always represented an important problem in Spain, mainly in the environment of aviation pilots, who constitute a very specialised staff.

Teaching and training have always been very expensive and the investment made by the Spanish state, must be as profitable as possible. This can be solved by contracts that stop this situation. Although, after the maximum staff stay times have passed, the problem cannot be avoided.

Generals: 52
Colonels and Lieutenant Colonels: 1,148
Commanders, Captains and Lieutenants: 2,174
Second Lieutenants and Brigades: 1,500
Sergeants and First Sergeants: 5,250
Troops: Lance Corporals, Professional Lance Corporals and Soldiers: 16,876
Civilian Personnel: 1,000
Total: approximately 28,000

In 2018, the EdA stated that in the last 10 years (2009–2018) the staff had lost more than 4,000 men/women, which meant that

28,000 military personnel had been reduced to 24,000. This decrease was expected to reach 20,000 by 2022.

It should be noted that of the 500 pilots in service in 2017, fifty-four had asked for leave to the benefit of the airlines.

12
THE SPANISH MILITARY AERONAUTICAL INDUSTRY

The Spanish aeronautical industry was born in the vicinity of Barcelona in 1916, when Pujol Comabella set up workshops that produced the Vendôme monoplane of French origin, fitted with a 50 HP Le Rhône engine. Later, the Monocoupe I and II models and the Pinguino Hedilla-Anzani 45 HP primary school monoplanes, inspired by the French Blériot XI model, were built there. After a few years and with a consolidated experience, this industry expanded and diversified, becoming Talleres Hereter, S.A. (in 1918). This new company built new models and obtained the confidence of the Spanish Army, which ordered six examples of the España type, an aircraft created by the Captain Eduardo Barrón, based on the French SPAD XIII model fitted with a Hispano-Suiza 180 HP engine.

At that time, the Hispano-Suiza and Elizalde firms were founded, and although at the beginning, they were dedicated exclusively to the production of automobiles, a little later they opened sections for the manufacture of aircraft engines, which very soon enjoyed a well-deserved international prestige. One of the most popular models was the 12-cylinder Hispano-Suiza with a power output of 650 HP.

It is important to mention that during this period, Construcciones Aeronáuticas (CASA) was created, an industry that has survived to the present day under different names.

All these industries and some others covered Spanish aeronautical needs for several decades until the outbreak of the Spanish Civil War (1936-1939). From that period onwards and due to wartime needs, both sides had to turn to foreign manufacturers: Germany, France, United Kingdom, Russia and Italy, preferably, who supplied them with all the aeronautical material they needed.

The post-war period found Spain totally ruined and with technologically outdated industries. The creation of the Ejército del Aire at the end of 1939 led the aeronautical authorities of the time to devote themselves, in some cases with very little success, to trying to build up these industries, always depending on imports, mainly from Germany, a country that was already immersed in the Second World War.

In the following, and given their links with the EdA, reference will be made briefly to the main ones: CASA and Hispano Aviación, S.A.

Construcciones Aeronáuticas, S.A. (CASA)

Established in 1923, it started its activity with the production under licence of Breguet XIX and Dornier Wal aircraft and later Junkers Ju.52, Heinkel He.111, Bücker Bü.131/133, Dornier Do.27 and CASA Northrop F-5 A/B.

In 1946, the CASA Project Office was created which from that time onwards, designed, developed and manufactured mainly transport aircraft such as the CASA C.201 Alcotán, C.202 Halcón, C.207 Azor, and the failed C.401 four-engine project. Few of them were successful, although these aircraft later led to the production of

CASA built a total of 73 Aviocars for the Spanish EdA between 1979 and 1977. It also manufactured aircraft of this type for several other countries. (Author's Photo)

CASA has produced several versions of the Aviocar C.212, both for military and civilian use. (Author's Photo)

In 1970, the CASA-Northrop F-5 designed for training was shown at various international trade fairs. (Author's Photo)

CASA workshops continue to produce C.235 aircraft. (Author's Collection)

The Puerto Real (Cádiz) factory is one of CASA's industrial complexes. (Author's Collection)

more successful aircraft such as the CASA C.212 Aviocar, the CASA-Nurtanio CN.235 and the more modern C.295. These three aircraft have achieved worldwide success, flying on all five continents.

It should be noted that in 1954, CASA was awarded the first contract for overhauling aircraft that the USAF had stationed in Europe, thus initiating a new international activity that lasted several decades, during which more than 6,000 aircraft and helicopters (C-47, C-54, F-86, F-100, T-28, T-33, F-101, F-102, F-105, C-130, S-16, HU-16, F-4, H-19 and Bell 47) passed through its workshops to be overhauled, repaired and modified.

This extensive experience allowed them to gain the expertise and technology necessary to produce the 70 CASA-Northrop F-5s for the EdA.

Later, because of a restructuring process in the Spanish aeronautical sector, CASA absorbed in 1972, La Hispano Aviación, S.A., and the following year, Empresa Nacional de Motores de Aviación, S.A. (ENMASA)

It also had extensive experience in the civil and astronautical fields. In the mid-1960s, it began working with other companies on the HFB-320 Hansa Jet project. Northrop Corporation became a shareholder in CASA in 1962, which led to the production of the F-5. In 1967, CASA was awarded its first space contract by ESRO. In 1969, collaboration began on *the Mercure Avions Marcel Dassault* aircraft and in 1971, on the Airbus A.320B project, leading to integration into the Airbus Industries Group.

Hispano Aviación, S.A. (HASA)

Hispano-Suiza, the predecessor of Hispano Aviación, S.A., was founded in Barcelona in March 1904 to manufacture automobiles. In 1911, the company's management opened a branch in Levallois, near Paris, which was followed by another branch in Bois-Colombes, also in the Paris region.

In 1915, it created his first engine, the 140 HP Hispano-Suiza with eight cylinders in V, of which 49,893 examples were produced in Spain, the United States, France, United Kingdom, Italy, and Japan.

Its first aircraft designed and built in its new factory located in Guadalajara was the Barrón-Hispano, a fighter that, although it met the requirements of the Spanish Government, given the needs of the African campaigns, was forgotten and replaced by the De Havilland DH.6. One hundred aircraft were contracted, fitted with the Hispano-Suiza 300 hp engine. Later, and also under licence, DH.9, Nieuport-Delage 52 C.1 were built. In the middle of the Civil War, a series of Polikarpov I-16 Rata and other models were built for the Republican side.

These two CASA built aircraft belong to the Argentine Naval Prefecture. (Author's Collection)

Indra has manufactured flight simulators for the EdA and also for other air forces around the world. (Author's Collection)

Once the war was over and under licence, it manufactured a hundred Fiat CR.35 aircraft, which was locally called HS.132L. It was at this time that the company changed its name to La Hispano Aviación, S.A. under which and under licence, it began to manufacture several series and versions of the Messerchmitt Bf.109G with English Rolls Royce engines. It also produced an all-Spanish training aircraft called the HS.42.

The accumulated experience allowed its technicians to develop several jet aircraft models, the most characteristic being the HA.200 Saeta, which flew for the first time in August 1955. It was powered by two Turboméca Marboré IIA, and a first series of 30 aircraft was produced in 1959. Internationally, the company signed a contract with the USAF for the overhaul of its T-33s.

On 13 June 1972, Hispano Aviación, with its projects, developments, and installations, became part of Construcciones Aeronáuticas, S.A.

This is briefly the history of the two main Spanish military (and civil) aeronautical industries. It should be mentioned that throughout more or less recent history, there have been some others who have contributed to the development of technologies that have allowed the main ones to work on the tasks that the Spanish Government and the EdA have entrusted to them.

Thus, after the creation of the EdA in 1939, other companies appeared that took over part of the industrial load that, due to their specialisation or small size, could not be carried out by the main companies. These were: Empresa Nacional de Motores de Aviación, S.A. (ENMASA), IBERAVIA (aviation projects), AERONÁUTICA INDUSTRIAL, S.A. (AISA), (small aircraft and helicopters), EMPRESA NACIONAL DE HÉLICES DE AVIACIÓN (ENHASA), AEROTÉCNICA (helicopters) and SANPERE PARACAÍDAS (parachutes). All of them were integrated or merged with CASA, or simply disappeared in the face of the push for more modern technologies.

BIBLIOGRAPHY

Andrade, John M., *Spanish and Portuguese Military Aviation* (MidLand Counties Publications, Leicester, England, 1977)

Arráez Cerdá, Juan, *La Máquina y la Historia, Nº 12: Hispano Aviación HA.200/220, SAETA Y SUPERSAETA* (Quirón Ediciones, Valladolid, 1997) (Spanish)

Ávila, Gonzalo & Yáñez, Roberto, *Alas Españolas: North American F-86F Sabre* (Reserva Anticipada Ediciones, Barcelona, 2000); *Alas Españolas: CASA C.2.111* (Reserva Anticipada Ediciones, Barcelona, 2003); *Alas Españolas: Hispano Aviación HA.200 Saeta* (Reserva Anticipada Ediciones, Barcelona, 2003) (Spanish)

Ávila Cruz, Gonzalo & González Serrano, José Luis, *Sabre, Comienzo de una Época* (Instituto de Historia y Cultura Aeronáutica, Madrid, 1988) (Spanish)

Base Aérea de Morón, 1941–1991 (Base Aérea de Morón, 1991) (Spanish)

Centro Fotográfico y Cartográfico del Ejército del Aire, 1920–1998 (CECAF, Madrid, 1998) (Spanish)

Cooper, Tom & Grandolini, Albert, *Africa @ War 33: Showdown in WESTERN SAHARA, Volume 1: Air Warfare Over the Last African Colony, 1945–1975* (Helion & Company Limited, Warwick, England, 2018)

Fresno, Carlos, *La Máquina y La Historia, Perfiles Aeronáuticos: Grumman SHU-16B Albatross Asw* (Ediciones Quirón, Valladolid) (Spanish)

García González, Alejandro, *Manises: El F-86F Sabre-Jet. Desde el Ala de Caza núm.1 (Una Época en las Fuerzas Aéreas Españolas)* (Ministerio de Defensa, Madrid, 1998)

González-Betes, Antonio, *Historia Gráfica de la Aviación Española* (Colegio de Ingenieros Aeronáuticos de España, Madrid) (Spanish)

Hebrero, Patricio & Ávila, Gonzalo, *Alas Españolas: Grumman HU-16 Albatross* (Reserva Anticipada Ediciones, Barcelona, 2002) (Spanish)

Herrera Alonso, Emilio, *Cuadernos de Historia Militar Nº 1: Aire, Agua, Arena y Fuego* (Quirón Ediciones Alcañiz-Fresno's Valladolid, 2001) (Spanish)

IHCA, Instituto de Historia y Cultura Aérea, *Fuerza Aérea Española* (Novatex Ediciones, Madrid, 1983); *Aviones Militares Españoles* (IHCA, Madrid, 1986) (Spanish)

Mellado Pérez, Rafael, *Academia General del Aire, Crónica de 50 años (1943–1993) Vol.1& Vol. 2* (Quirón Ediciones, Valladolid, 1993) (Spanish)

Pando, Juan Carlos, *Alas Españolas: CASA CN.235M* (Reserva Anticipada Ediciones, Barcelona, 2000); *Alas Españolas: CASA C.101 Aviojet* (Reserva Anticipada Ediciones, Barcelona, 2001); *Alas Españolas: CASA C.212 Aviocar* (Reserva Anticipada Ediciones, Barcelona, 2003) (Spanish)

Paniagua Magán, Pedro Miguel, *Monográfico Aéreo: H.A. 1112 – M1L Buchón* (Monográficos Aéreos, Madrid, 1995) (Spanish)

Patrulla Águila, Patrulla Acrobática de España (Patrulla Águila, AGA, Murcia, 1995) (Spanish)

Pecker, Beatriz y Pérez-Grange, Carlos, *Crónica de la Aviación Española* (SILEX, Madrid, 1983) (Spanish)

Redón, Pere (ed.), *OPERACIÓN ICARO. El Ejército del Aire en Bosnia–Herzegovina, Noviembre 1994 – Diciembre 1995* (Pedro Redón, Barcelona, 1996) (Spanish)

Redón, Pere, *Alas Españolas: North American T-6 Texan* (Reserva Anticipada Ediciones, Barcelona, 1999); *Alas Españolas: McDonnell Douglas A/F-18, F-18B Hornet* (Reserva Anticipada Ediciones, Barcelona, 2000); *Alas Españolas: Lockheed F-104G Starfighter* (Reserva Anticipada Ediciones, Barcelona, 2000); *Alas Españolas: Marcel Dassault, Mirage III* (Reserva Anticipada Ediciones, Barcelona, 2002); *Alas Españolas: Lockheed C-130 Hercules* (Reserva Anticipada Ediciones, Barcelona, 2002); *Alas Españolas: McDonnell Douglas F-4C, RF-4C Phantom II* (Reserva Anticipada Ediciones, Barcelona, 2004) (Spanish)

Salas Larrazábal, Jesús, *De la Tela al Titanio* (Espasa-Calpe, Madrid, 1983); *LA HISPANO AVIACIÓN, Proyectos HA.100, 200 y 300* (Ministerio de Defensa, Madrid, 1999) (Spanish)

Sanz Peñalver, Juan, *Reseña Histórica de la Base Aérea de Morón – Ala 11, 1941–2001* (Ministerio de Defensa, Madrid, 2002) (Spanish)

Seminario de Estudios Históricos Aeronáuticos, *Grandes Vuelos de la Aviación Española* (Espasa-Calpe, Madrid, 1983) (Spanish)

Terol, José, *El Álbum del Piloto* (Cuartel General del Ejército del Aire, Madrid, 1992) (Spanish)

Yáñez, Roberto & Ávila, Gonzalo, *Alas Españolas: Canadair CL-215 – CL-215T* (Reserva Anticipada Ediciones, Barcelona, 2000); *Alas Españolas: Dassault Mirage F-1* (Reserva Anticipada Ediciones, Barcelona, 2000) (Spanish)

ZARAGOZA, Base Aérea (Base Aérea de Zaragoza, 1987) (Spanish)

Magazines:

Revista de Aeronáutica y Astronáutica (Ministerio de Defensa, Ejército del Aire, Madrid) (Spanish)

Aeroplano, Revista de Historia Aeronáutica (IHCA, Ministerio de Defensa, Ejército del Aire, Madrid) (Spanish)

Internet:

https://ejercitodelaire.defensa.gob.es

ABOUT THE AUTHOR

Pere Redón-Trabal is a journalist and writer specialising in military subjects, especially aviation. He has collaborated in numerous technical and informative magazines and created and edited the legendary Spanish aviation magazine '*Air Sonic*'. In recent years, he has authored and edited the following books (written in Spanish) '*Operation Icaro*', '*The Army in peacekeeping missions*', '*The sea and peace*', '*The Guardia Civil in peacekeeping missions*', '*A History of FAMET*', '*Maritime Service of the Guardia Civil*', '*Sicut in coelo et in terra*', and '*Logistics Brigade, proud to serve*'. He has a large archive and library, as well as an extensive photographic collection, which he uses in his work.